Advances in Nonconventional Machining Processes

Edited by

Suneev Anil Bansal

Department of Mechanical Engineering
Maharaja Agrasen University
India

Advances in Nonconventional Machining Processes

Editor: Suneev Anil Bansal

ISBN (Online): 978-981-14-8365-3

ISBN (Print): 978-981-14-8363-9

ISBN (Paperback): 978-981-14-8364-6

need for a court order if at any point you breach any terms of this License Agreement. In no event will any delay or failure by Bentham Science Publishers in enforcing your compliance with this License Agreement constitute a waiver of any of its rights.

3. You acknowledge that you have read this License Agreement, and agree to be bound by its terms and conditions. To the extent that any other terms and conditions presented on any website of Bentham Science Publishers conflict with, or are inconsistent with, the terms and conditions set out in this License Agreement, you acknowledge that the terms and conditions set out in this License Agreement shall prevail.

Bentham Science Publishers Pte. Ltd.
80 Robinson Road #02-00
Singapore 068898
Singapore
Email: subscriptions@benthamscience.net

CONTENTS

PREFACE

Machining is one of the important manufacturing processes. Both conventional and non-conventional machining processes play a vital role in the industry.

In the modern era of manufacturing, Non-Conventional machining methods are quite popular due to various advantages such as high accuracy, excellent surface finish, less tool wear, much quieter operations *etc*. Moreover, new age and novel materials are sometimes hard to machine with traditional machining processes due of their high hardness, toughness, brittleness *etc*. Non-Conventional machining methods are now being commonly and frequently used in the industry. Present book 'Advances in Non-Conventional Machining Processes' aimed to through light on recent development in these methods.

The major aim of this book is to provide the readers detailed information regarding various advances in the field of Non-Conventional machining. In addition to provide cutting edge information, the book also describes the level of modern technology adopted by various industries.

Many Authors have contributed their ideas on various aspects of Non-Conventional machining processes. The prime focus of all the authors is to provide the latest and exact information regarding various topics. In total, nine chapters have been contributed to this book that touches various fields of Non-Conventional machining such as Abrasive Jet machining, Electrical Discharge Machining, Thermal machining *etc*. Keeping in view, the trend of this topic for researchers well cited references are given at the end of each chapter for further information of the topic. Relation between Non-Conventional machining and Artificial intelligence and their relationship with modern industry has also been focused in one of the chapters.

An attempt has been made to provide quality research data and this book is intended both for academia and industry. The former refers to pre & post graduate students along with PhD students, who are working on the field of manufacturing while the later refers to the industries in which R& D departments are constantly working towards the betterment of the machining process to produce sophisticated products.

Suneev Anil Bansal
Department of Mechanical Engineering
Maharaja Agrasen University
India

List of Contributors

Amit Handa	I.K. Gujral Punjab technical University, Kapurthala, India
Amrinder Pal Singh	Department of Mechanical engineering, UIET, Panjab University Chandigarh, India
Ankitmani Tripathi	Mechanical Engineering Department, SRM University, Delhi, India
Anupam Thakur	Department of Mechanical and Automation Engineering, Maharaja Agrasen Institute of Technology, Delhi, India
Anurag Thakur	Department of Mechanical Engineering, Alakh Prakash Goyal shimla University Shimla, Himachal Pradesh, India
Arminder Singh Walia	Mechanical Engineering Department, Thapar Institute of Engineering & Technology, Patiala, India
Balwinder Singh	Centre for Development of Advanced Computing, Mohali, India
Bhuvnesh Kumar	Mechanical Engineering Department, SRM University NCR Haryana, Delhi, India
Chander Prakash	School of Mechanical Engineering (SME) Lovely professional University, Jalandhar, India
Hitesh Pahuja	Centre for Development of Advanced Computing, Mohali, Punjab, India
Janender Kumar	Department of Mechanical Engineering, Maharaja Agrasen University, India
Kamaljit Singh	Chandigarh Engineering College Landran Mohali, India
Mamta	Department of Mechanical Engineering, Maharaja Agrasen University, Baddi, Solan, India
Munish Mehta	School of mechanical engineering Lovely Professional University, India
P.K Khosla	Centre for Development of Advanced Computing, Mohali, Punjab, India
Ramakant Rana	Department of Mechanical and Automation Engineering, Maharaja Agrasen Institute of Technology, Delhi, India
Sachin Mohal	Department of Mechanical Engineering, Chandigarh Engineering College Landran, Mohali, India
Saurabh Chaitanya	Department of Mechanical Engineering, Chandigarh Engineering College Landran, Mohali, India
Suneev Anil Bansal	Department of Mechanical Engineering, Maharaja Agrasen University, India
Vineet Srivastava	Mechanical Engineering Department, Thapar Institute of Engineering & Technology, Patiala, India
Virat Khanna	Department of Mechanical Engineering, Maharaja Agrasen University, Baddi, Solan, India
Vivek Jain	Mechanical Engineering Department, Thapar Institute of Engineering & Technology, Patiala, India

CHAPTER 1

Ultrasonic Machining Process - A Review

Janender Kumar[1,*], Amrinder Pal Singh[2], Anurag Thakur[3] and Munish Mehta[4]

[1] *Department of Mechanical Engineering, Maharaja Agrasen University, Baddi, Solan, India*

[2] *Department of Mechanical engineering UIET, Panjab University Chandigarh, India*

[3] *Department of Mechanical and Automation Engineering, Maharaja Agrasen Institute of Technology, Delhi, India*

[4] *School of Mechanical Engineering, Lovely Professional University, Phagwara, Punjab, India*

Abstract: The utilization of advanced materials like ceramics, composites *etc.* has increased these days and simultaneously machining of these hard, brittle and costly materials is very challenging. Industries are relying on non-conventional machining processes because conventional machining processes have many limitations to machine hard and brittle materials. The birth of modern machining processes like Ultrasonic Machining (USM) has taken place due to these limitations in conventional machining. In USM, material is removed due to action of abrasive grains and vibrations cutting tool during machining. Up to some extent, it overcomes the problems of conventional machining. Besides USM has some drawbacks *e.g.* low material removal rate, high wear rate of tool. As it tools continuously strikes on the workpiece, there is oversize in produced cavities and a limit of depth in drilling operation. Optimization of process parameters can improve all these problems. The literature survey study has shown, in drilling and milling operations, a hybrid machining process named Rotary Ultrasonic Machining (RUM) gives tremendous results as compared to ultrasonic machining alone. In recent years, it is gaining popularity due to its intelligent machining in industries. The present paper confers about the technique of parametric optimization of ultrasonic machining and future aspects of rotary ultrasonic machining processes.

Keywords: Abrasive particles shape and sizes, Ceramics and composites, Hard and brittle material, Hybrid machining, Machining parameters, Rotary ultrasonic machining, Ultrasonic machining.

* **Corresponding author Janender Kumar:** Department of Mechanical Engineering, Maharaja Agrasen University, Baddi, Solan, India; E-mail: janyo_ishaan@yahoo.com

INTRODUCTION

In the modern world, due to technological advancements, the needs of industry are continuously changing. Latest and advanced materials, along with their machining, are the requirement of the manufacturing sector. Research is a continuous process and it's the demand/necessity of growing industries. Truly speaking, change for better is necessary. In earlier days machining of hard and brittle material was a tough job for engineers and was a challenge for the research and design team as well. Conventional machining was used for many years. Newer processes of machining were developed and named as modern machining methods/non-conventional machining. Recent development of these processes opened the way of machining hard and brittle materials. In conventional machining processes, there is direct contact between tool and workpiece and material removed in form of chips *i.e.* at macroscopic level, whereas in non-conventional machining there is no direct contact between the tool and work piece. In this, the material is removed at microscopic level (dust/powder form). In conventional machining, due to physical contact tools, wear rate is high as compared to non-conventional machining. The increasing utility of hard and brittle materials in recent times has forced the researchers to develop new and advanced machining processes. The materials, difficult to machine through the conventional machining, are machined by unconventional machining processes. The main idea behind the development of newer methods was the replacement of conventional machining methods by more efficient, quicker methods which could achieve accuracies in machining. Ultrasonic machining is one such development in the field of non-conventional methods. In this process, slurry of small abrasive particles is forced against the work by an axially vibrating tool, removing the workpiece material in the form of small particles. The abrasive grain particles may be of silicon carbide, boron carbide and diamond dust. To facilitate/provide high precision machining, when vibration is introduced in a tool workpiece or working medium then it is called vibration assisted machining [13]. In vibration assisted turning, cutting forces reduced up to 40% as compared to conventional machining, hence improving tool life. Less stresses are induced in work and quality of machined surface is improved [15]. Rotary ultrasonic machining is one of the latest processes which are used in industries to overcome the limitations of ultrasonic machining. Higher metal removal rate is observed during drilling operation through rotary ultrasonic machining as compared to ultrasonic machining alone [23].

LITERATURE REVIEW

Kumar and Khamba evaluated the machining performance of ultrasonic machining on titanium-based workpiece. The investigation was carried out for material removal rate, tool wear rate and to know the surface roughness of the workpiece. Taguchi's method was adopted for performing several experiments, the data was analysed by using ANOVA and optimised parameters were selected which can enhance the performance of ultrasonic machining [1].

B. Lauwers *et al.* investigated the material removal rate, surface quality, surface cracks in ultrasonic assisted grinding of ZrO_2. The paper showed that cutting forces and tool wear were reduced under the influence of vibration during machining. Hence productivity increased [2].

Cong *et al.* presented the method of measuring vibration amplitude during rotary ultrasonic machining. In this paper, it was concluded that vibration amplitude varies for different conditions. When the dial indicator method was used for measuring vibration amplitude and compared with the novel method, results were the same. But when it was checked with or without rotary ultrasonic machining, using tools with different specifications, results were different [3].

Heise *et al.* investigated the ultrasonic assisted drilling operation on granite and stone, which is a hybrid machining process. The cutting forces and torques was reduced up to 20% when the ultrasonic process applied. That could be achieved by using robots instead of machining centres [4].

Kei-Lin and Chung-Chen presented the paper on rotary ultrasonic machining, in which glass was utilised for milling operation. The efficiency of operation was evaluated based on surface roughness parameter. It was seen that the surface quality of ultrasonic milling was very poor as compared to non-ultrasonic milling [5].

Liu *et al.* developed a model for cutting forces during rotary ultrasonic machining. Cutting forces play a very important role during machining of brittle material. A relation between cutting force and input variables like spindle speed, feed rate, amplitude of ultrasonic vibration and abrasive grain particles was developed based on the cutting forces model. The various assumptions were made during research analysis like abrasive particles' shape and its material. Diamond abrasive particles, of octahedron shape, were utilized. The developed model gave information for cutting forces *i.e.* forces increased as input variables parameters increased and decreased as input variable parameters decreased [6].

Rasheed compared the results of micro-holes produced by micro-EDM and laser beam machining. The machining parameters like material removal rate, taper in hole, entrance and exit diameters of produced holes, concentricity of the produced holes was compared. It was concluded that when surface quality of the workpiece is not so important then for high material removal rate, laser beam machining can be utilised. If surface quality is needed, then micro-EDM can be used but material removal rate will be reduced. So, a hybrid process can be made to take the advantages of both processes [7].

Das *et al.* investigated the surface roughness of hexagonal shaped hole and material removal rate in Zir-Conia bio-ceramic material after ultrasonic machining. With control parameters, optimised results were obtained by utilising genetic algorithm technique. In the conducted experiment, with optimised parameters, material removal rate of and surface roughness of 0.2365g/min and 0.58-micron meter were obtained respectively [8].

Lian *et al.* performed several experiments to analyse the surface roughness of A16061 material by micro-milling with or without the assistance of ultrasonic vibrations. It was observed that better surface finish was obtained in A16061 with ultrasonic vibration assisted machining. Better results were obtained when appropriate ultrasonic vibration amplitude was selected [9].

Cong *et al.* provided an explanation about surface roughness of hole, in ultrasonic machining, at entrance and exit. The paper considered hypotheses and their testing. The experiment was performed on a stainless-steel workpiece. It concluded that the reason for the difference of surface roughness at entrance and exit of the workpiece depends upon the abrasive bonded part of the tool movement in the workpiece through the entire length [10].

Selvakumar and Arulshri carried out the work to optimise the fixture layout using genetic algorithm and combination of genetic algorithm and artificial neural network approaches. The purpose of investigation was to reduce the deformation of the workpiece after clamping on fixture and machining force during operation. Finally, results of genetic algorithm and artificial neural network approach were compared. It proved that the combination of genetic algorithm and artificial neural network approach was better than genetic algorithm [11].

Yang *et al.* performed study on ultrasonic lapping of gears. It is a better process for tooth surface quality of gears as compared to conventional methods of lapping. The purpose of this paper was to finalise the methodology of ultrasonic lapping design for hypoid gear. After experiment, it was concluded that ultrasonic lapping of gear transmission quality was better than conventional lapping method [12].

Feucht *et al.* presented the machining results of hard and brittle materials or advanced materials such as ceramics and fiber- reinforced materials which are used in medical and aerospace industries using ultrasonic assisted machining. Fibre reinforced material was inhomogeneous in nature. So, it remains difficult to machine this material. But when vibration was produced by transducer superimposed on the spindle of conventional machining process, the cutting forces and tool wear rate were reduced by 20% to 30% as tool periodically lifted from workpiece or job [13].

Muhammad *et al.* discussed about hybrid machining technique in which ultrasonic assisted turning and hot machining technique were used in combination to machine titanium alloys. As a result, cutting forces was reduced and surface roughness was improved drastically in titanium alloys. In hot ultrasonic assisted turning, 50% excess improvement was noticed in surface roughness [14].

Kumar *et al.* proposed vibration assisted machining in cutting hard and brittle material. Vibration can be given to tools, workpiece and working medium. During ultrasonic assisted turning, the process parameters improved like 40% to 45% reduction in cutting forces and average cutting forces dropped to 40% as compared to conventional machining. Tool wear rate was reduced, and surface roughness was improved in the workpiece [15].

Bhosale *et al.* invented the effect of process parameters on alumina-zirconia ceramic composite on material removal rate, tool wear rate and surface topography. The process was analysed through the effect of amplitude, slurry of abrasive particles with carrying fluid *etc.* The research showed that tool wear rate tends to increase when hard and coarse abrasive was used. High tool wear rate occurred when boron carbide abrasive is utilized as compared to silicon carbide abrasive for similar machining conditions. The study concluded that slurry concentration has less effect on tool wear rate while amplitude increases surface roughness [16].

Silberschmidta *et al.* performed the experiment on ultrasonic assisted turning in which a combination of cutting parameters were tried on machines for different materials like copper, alloys *etc.* The cutting forces reduced during machining and surface roughness improved due to on ultrasonic assisted turning. The result obtained, for inconel 718, was better than conventional turning for surface roughness. The Ra value in case of conventional turning was 1.179-micron meter as compared to 0.505-micron meter in case of ultrasonic assisted turning [17].

Goswami and Chakraborty utilized two techniques for parametric optimization of ultrasonic machining processes. These were gravitational research and firework logarithm. Authors developed the best optimal results like high material rate as compared to gravitational research [18].

Agarwal investigated the ultrasonic machining of glass material. The study focused on mechanisms of material removal and material removal rate. It showed that as grit size increases, material removal rate also increases. The shocking force and material removal rate relation was developed. The analysis represented that increasing static load amplitude of tool tip, the material removal rate can be increased. On the other side, with increase in grit size, material removal rate was reduced because debris of the workpiece did not allow the tool to vibrate. Study revealed that proper controlling parameters can improve material removal rate only [19].

Ning discussed the importance of CFRP composites used in aerospace and aircraft industries due to its superior properties. The two drilling methods were compared, rotary ultrasonic machining and grinding. The cutting parameters like cutting forces, surface finish, torque, hole diameter and material removal rate were taken into consideration in the study. The comparison showed that rotary ultrasonic machining performed better in drilling of CFRP than grinding [20].

Wang *et al.* studied the problem of tearing defects at hole exit during drilling a hole in C/Sic composite material. The new tool was developed and compared with common drill. The result at hole exit was better with using compound step-taper drill. Problem of tearing defects got reduced due to using this drill. The following machining parameters were maintained like a feed rate of 2 mm/min. and spindle speed of 2000 rpm. The results showed that tearing size reduced up to 30% using compound step- taper drill. Defects produced during exit after drilling were burr, edge chipping *etc* [21].

Aminiet *et al.* studied tribological properties during ultrasonic vibration assisted turning process. The cutting parameters like friction coefficient and wear rate were considered for evaluation. ANOVA results indicated that effectiveness of 32.32% was observed in respect of friction coefficient [22].

Singh and Singhal reviewed the non-conventional hybrid machining process *i.e.* rotary ultrasonic machining. As in static ultrasonic machining, abrasive particles insert between tool and workpiece continuously during machining. The vibrating tool strikes on the workpiece and material removal take place. In this process, there was no direct contact between tool and workpiece. So, lesser material removal took place. The result of machining was not as per specifications. Due to which defects produced in drilling such as oversized holes, eccentric holes and erosion of wall surfaces *etc.* in rotary ultrasonic machining, conventional diamond grinding and ultrasonic machining combined together. Hence higher material removal rate produced as compared to conventional grinding and ultrasonic machining. In rotary ultrasonic machining, rotation was given to vibrating drilling tools and abrasive particles of diamond merged on tool tip during drilling operation. This process gave higher material removal, improved surface roughness and accuracy of machined holes than conventional grinding and static ultrasonic machining [23].

Zhen *et al.* carried out research on ceramic matrix composites. In the study, it was observed that rotary ultrasonic machining increased the efficiency of machining 5.8 times more and surface quality improved by 54% as compared to conventional milling [24].

Ning *et al.* presented the comparison between rotary ultrasonic surface machining and conventional surface grinding of CFRP composite material. The kinematics motion of abrasive particles was studied, and its comparison was also carried out. During analysis of kinematic motion of abrasive particles, it was assumed that abrasive particles are spherical in shape with the same diameter. The shape remains the same during machining. The experiment results indicated that rotary ultrasonic surface machining produced lower values of infeed cutting forces, axial cutting force and also torque. Rotary ultrasonic surface machining besides this gave larger surface roughness as compared to conventional surface grinding. The reason for this was vertically ultrasonic vibration [25].

Feng *et al.* discussed the tearing defect at hole exit during drilling. The experiment showed that the defect was due to large thrust forces. Rotary ultrasonic machining reduces the thrust force and 60% reduction has been observed in tearing defects. Experiment results indicated that it reduces the thrust force approximately 55% in the study. The grain size has much more influence on thrust force and it was decreasing up to some extent [26].

Wang *et al.* studied the tool wear mechanism and its relation to material removal rate. In the trial, three different material tools are used, named 304 stainless steel, 1045 carbon steel and tungsten. The influence of tool wear shows low MRR and inaccuracy in machining. During machining, it was observed that 304 stainless steel tools are best for micro USM. The tool wear and abrasive particle wear was low which helps to remove more material. The 304 stainless steel tool has low internal surface hardness as compared to external surface. The combined surface material (soft & hard) gave insurance of low tool wear and particle wear [27].

Wang *et al.* studied the potential of ultrasonic machining and investigated on the shape of abrasive particles. In the research, two different shaped abrasive particles were taken. The shapes were cubicle and spherical which were filled with silicon carbide and aluminum oxide materials. The result suggested that hard abrasive particles with spherical shape (Al_2O_3) resisted the tool wear in a nice way as shown in Fig. (**1**). This improves the material removal efficiency during ultrasonic machining and also its performance [28].

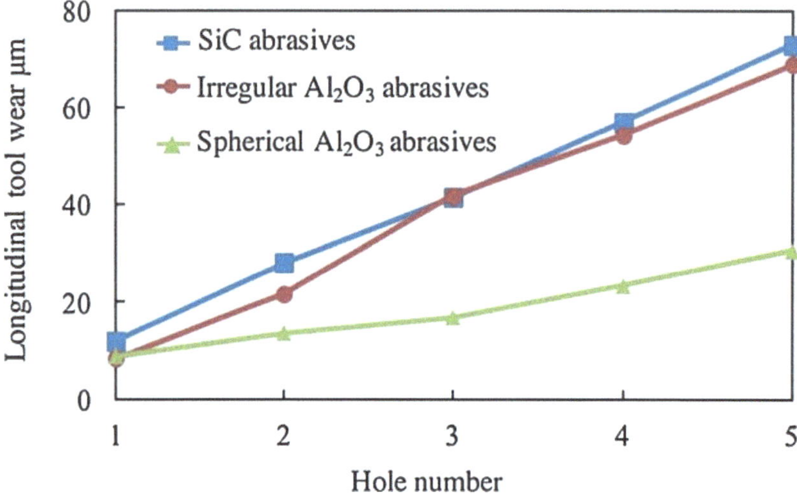

Fig. (1). Trend of tool wear rate with different abrasive particles used during drilling.

Wang *et al.* presented the cutting force and material removal mechanism in rotary ultrasonic machining. The defect produced drilling briefed like edge chipping, tearing defect and subsurface damage. The damage reduction technique and optimization of processing parameters were also described. During machining, damages was produced at hole exit. This could be reduced if cutting forces were less than critical cutting forces. By optimizing process parameters, damage during machining can be reduced [29].

Wang *et al.* investigated the critical cutting forces in rotary ultrasonic machining. The cutting forces were analyzed to develop a novel index which was required for design and manufacturing of an ultrasonic machine tool. In the paper, the effect of cutting force was analysed on processing parameters and ultrasonic amplitude. Also, critical cutting forces effect on rotary ultrasonic m/c tools were analysed & method of improvement discussed [30].

Popli & Gupta investigated edge chipping problems during drilling with rotary ultrasonic machining. An experiment was designed to remove the edges produced during drilling at hole exit by increasing the support length. It helped in reducing the edge chipping size up to 50%. The relation between chipping size and support length was given by chipping size proportional to 1/support length [31].

Mikhailova *et al.* presented the results of drilling techniques of conventional drilling and ultrasonic assisted drilling in marble. Experiment's results are in favor of ultrasonic assisted drilling. It was observed that jobbler drill bit is preferable for drilling in marble also, but the feed rate has certain limits [32].

Wang *et al.* developed a surface model depending upon kinematics analysis and fracture mechanics of brittle materials. This model helped in predicting that at which parameters surface roughness will increase or decrease. In this paper when spindle speed was 2000 rpm, surface roughness was very low and surface topography was better but when spindle speed was between 2500-3000 rpm, surface roughness was high [33].

Wang *et al.* investigated the problem of high cutting forces, high surface roughness during machining of hard and brittle material like CFRP composite. Rotary ultrasonic surface machining, with vertical ultrasonic vibrations up to some extent, reduced the cutting forces but produced higher surface roughness than conventional surface grinding. To improve this parameter, experiment on rotary ultrasonic surface machining with horizontal ultrasonic vibrations was performed. This helped in reducing cutting forces and improved surface roughness than conventional surface grinding [34].

Wang *et al.* pointed out the material removal mechanism in rotary ultrasonic surface machining of CFRP composite. For this single abrasive scratching test has been conducted. At small scratching depth with intender of single diamond grain size whose point angle was 100°, ductile material removal mode mechanism was observed, and brittle removal mode was observed when scratching depth was large. In ductile material removal mode, surface observed were without crack whereas in brittle material removal mode, cracks were observed during machining. Similarly, with less scratching depth, force was less in ductile mode and in brittle material removal mode forces was large [35].

During study of research papers, it was observed that rotary ultrasonic process is superior to ultrasonic machining.

CONVENTIONAL MACHINING *VERSUS* NON-CONVENTIONAL MACHINING

Conventional machining is a traditional machining process in which the tool is in direct contact with the workpiece during machining and has limitations to machine certain material and to produce complex shapes whereas non-conventional machining has no direct contact between tool and workpiece. It has no limitation to machine any material.

ULTRASONIC MACHINING *VERSUS* ROTARY ULTRASONIC MACHINING

Ultrasonic machining is a machining process in which shaped tool conversely to the required cavity is axially vibrated and fed towards the work piece with controlled parameters [36]. Slurry is inserted between tool and workpiece which plays the role of cutting. The schematic diagram of ultrasonic machining with basic elements and working principle are shown in Figs. (**2** and **3**).

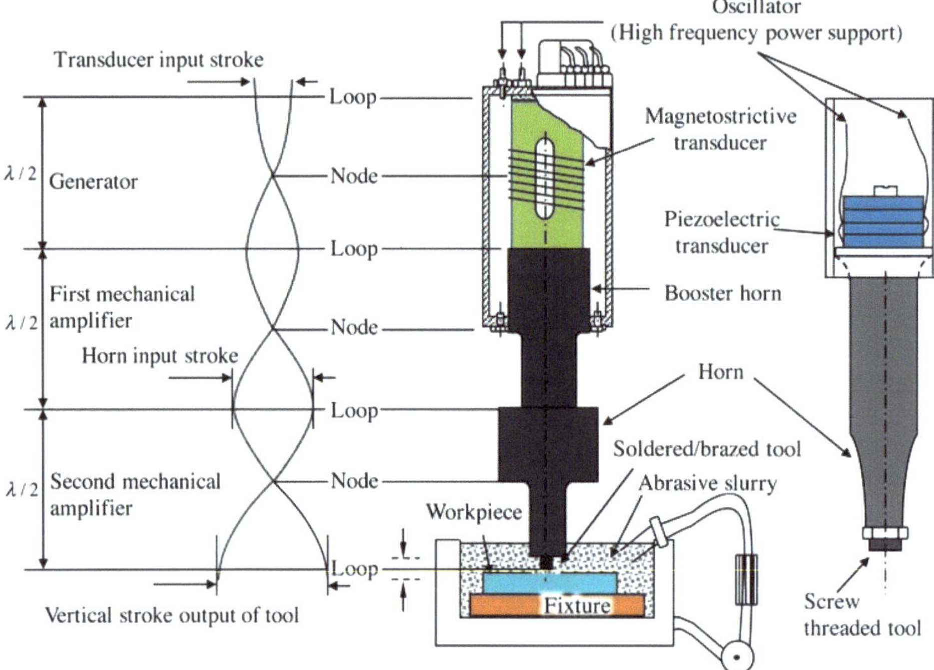

Fig. (2). Schematic diagram of Ultrasonic Machining set up [28].

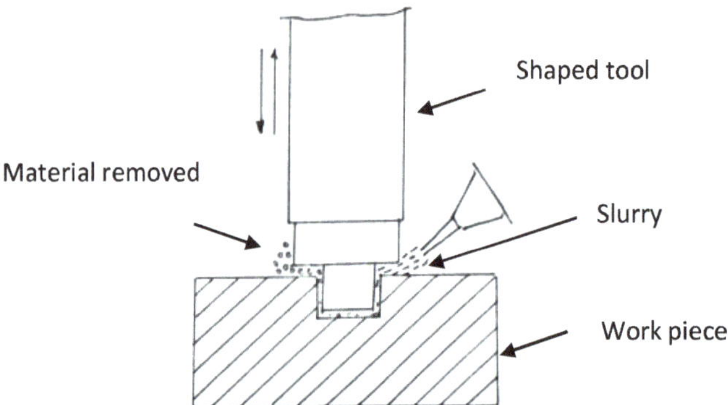

Fig. (3). Working Principle of Ultrasonic Machining Process.

Whereas in rotary ultrasonic machining, drill rotates continuously and is also having axial vibrations [37]. No separate slurry is fed between tool and workpiece. Cutting tool is specially designed in which abrasive grain particles are bonded on the tool body as shown in Fig. (**4**) [38]. In rotary ultrasonic machining abrasive particles are directly bonded on drill tools. Then rotary and vibrated drill provides a fast, accurate and high-quality hole in a variety of materials like glass, ceramic materials *etc* [39].

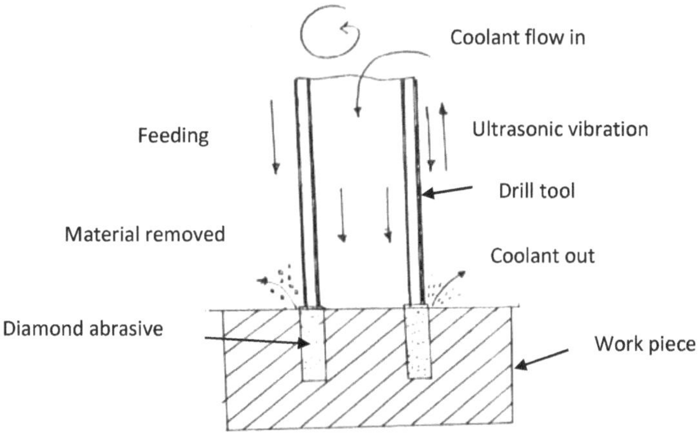

Fig. (4). Working Principle of Rotary Ultrasonic Machining Process.

Rotary ultrasonic machining is a hybrid machining process which combines ultrasonic machining and conventional grinding together for getting better machining results [23].

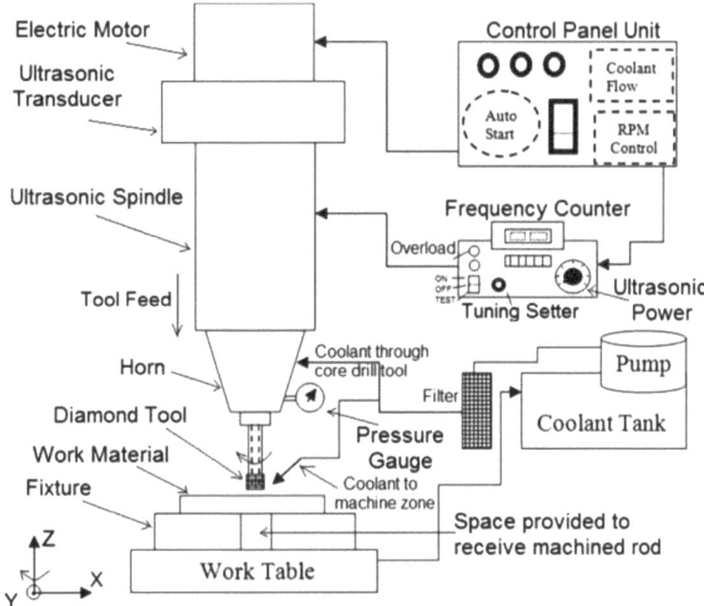

Fig. (5). Schematic diagram of Rotary Ultrasonic Machining set up [23].

The study reveals that only the controlled parameters of ultrasonic machining can increase the machining efficiency. The parameters are amplitude of vibrations, abrasive grain particles shape and size [25, 28], effect of applied static load *etc.* These controlled parameters can reduce the cutting forces for better performance of process [6, 13 - 15, 17]. Rotary ultrasonic machining is the up gradation for static ultrasonic machining [40]. The schematic diagram of rotary ultrasonic machining is shown in Fig. (**5**).

CONCLUSIONS

1. Ultrasonic machining has overcome the conventional machining problems in which the tool remains in direct contact with the workpiece whereas in ultrasonic machining, tool is periodically lifted from the workpiece or job, hence reduced tool wear rate.

2. It can machine hard and brittle material which has a restriction in conventional machining process

3. Ultrasonic machining features get improved when combined with other conventional and non-conventional machining.

4. Firework algorithm, a mathematical approach for parameters optimisation of ultrasonic machining, can be implemented.

5. Vibration assisted machining eliminates the use of cutting fluid or coolant. During machining, periodic work-tool separation maintains the gap which helps in dissipating heat from the cutting zone.

6. It helps in reducing environmental hazards which was due to coolant and improves the machining efficiency.

7. During literature review, it is observed that rotary ultrasonic machining has been utilized for drilling operation which shows good results like high material removal rate, low tool wear rate and good surface finish as compared to static ultrasonic machining drilled components.

8. In hard and brittle materials, rotary ultrasonic machining can be preferred for drilling operation instead of static ultrasonic machining for better results.

9. Research shows that in recent years, rotary ultrasonic machining is gaining popularity due to its intelligent machining of hard and brittle material. In aerospace industries, a gang of drills are utilized to create numbers of holes at same time with this process.

10. For better results spherical shape abrasive particles can be utilised rather than sharp edges particles.

11. The consumption of abrasive particles can be reduced in rotary ultrasonic machining as it is bonded on tool surface whereas in ultrasonic machining, it inserts separately.

12. Finally, static ultrasonic and rotary ultrasonic machining has solved the problem of machining the hard and brittle material which is very superior in quality and also costly.

FUTURE ADVANCEMENTS

During study of research papers, it was noticed that rotary ultrasonic machining was utilised for drilling and surface milling operations in hard and brittle materials only. It showed its restrictions in other operations. Even it doesn't talk about other drilling related operations like spot facing, counter boring, milling cum drilling combined and tapping also. There may be certain reasons behind this but left some area for further research work.

CONSENT FOR PUBLICATION

Not applicable.

CONFLICT OF INTEREST

The authors confirm that this chapter content has no conflict of interest.

ACKNOWLEDGEMENTS

The authors would like to express their sincere thanks to the editor and anonymous reviewers for their time and valuable suggestions.

REFERENCES

[1]   J. Kumar, and J.S. Khamba, "Multi-response optimisation in ultrasonic machining of titanium using Taguchi's approach and utility concept", *Int. J. Manuf. Res.,* vol. 5, no. 2, pp. 139-160, 2010.
[http://dx.doi.org/10.1504/IJMR.2010.031629]

[2]   B. Lauwers, F. Bleicher, P. Ten Haaf, and J. Loenders, "Investigation of the process-material interaction in ultrasonic assisted grinding of Zro2 based ceramic materials", *Proceedings of 4th CIRP Internationl on high performance cutting,* 2010pp. 1-7

[3]   W.L. Cong, Z.J. Pei, N. Mohanty, E. Van Vleet, and C. Treadwell, "Vibration amplitude in rotary ultrasonic machining: a novel measurement method and effects of process variables", *J. Manuf. Sci. Eng.,* vol. 133, pp. 034501-034505, 2011.
[http://dx.doi.org/10.1115/1.4004133]

[4]   U. Heise, "l,R.Eisseler, R.Eber, J.Teiefel and M.Huang,"Ultrasonic-assisted machining of stone", *Prod.Eng.Res.Devel,* vol. 5, pp. 587-594, 2011.
[http://dx.doi.org/10.1007/s11740-011-0330-1]

[5]   "Rotary ultrasonic-assisted milling of brittle materials", *Trans. Nonferrous Met. Soc. China,* vol. 22, pp. 793-800, 2012.

[http://dx.doi.org/10.1016/S1003-6326(12)61806-8]

[6]   D. Liu, W.L. Cong, Z.J. Pei, and Y. Jun Tang, "A cutting force model for rotary ultrasonic machining of brittle materials", *Int. J. Mach. Tools Manuf.,* vol. 52, pp. 77-84, 2012.
[http://dx.doi.org/10.1016/j.ijmachtools.2011.09.006]

[7]   M.S. Rasheed, "Comparison of micro-holes produced by micro-EDM with laser", In: *International journal of science and modern engineering* vol. 1. , 2013, no. 3, pp. 14-18.

[8]   S. Das, B. Doloi, and B. Bhattacharyya, "Optimisation of ultrasonic machining of zirconia bio-ceramics using genetic algorithm", *International journal of manufacturing technology and management,* vol. 27, no. 4/5/6, pp. 186-197, 2013.

[9]   H. Lian, Z. Guo, Z. Huang, Y. Tang, and J. Song, "Experimental research of A16061 on ultrasonic vibration assisted micro-milling", *Procedia CIRP,* vol. 6, pp. 561-564, 2013.
[http://dx.doi.org/10.1016/j.procir.2013.03.056]

[10]   W.L. Cong, Z.J. Pei, T.W. Deines, P.F. Zhang, and C. Treadwell, "Surface roughness in rotary ultrasonic machining: Hypotheses and their testing *via* experiments and simulations", *Int. J. Manuf. Res.,* vol. 8, no. 4, pp. 378-393, 2013.
[http://dx.doi.org/10.1504/IJMR.2013.057748]

[11]   S. Selvakumar, and K.P. Arulshri, "Machining fixture layout optimisation using genetic algorithm and artificial neural network", *Int. J. Manuf. Res.,* vol. 8, no. 2, pp. 171-195, 2013.
[http://dx.doi.org/10.1504/IJMR.2013.053286]

[12]   "J,J,Yang,H.Zhang,X.Z.Deng and B.Y.Wei,"Ultrasonic lapping of hypoid gear: system design and experiments", *Mechanism Mach. Theory,* vol. 65, pp. 71-78, 2013.
[http://dx.doi.org/10.1016/j.mechmachtheory.2013.03.002]

[13]   F. Feucht, J. Ketelaer, and A. Wolff, "Masahiko Mori, Makoto Fujishima, "latest machining Technologies of Hard-to-cut Materials by Ultrasonic Machine Tool", *6th CIRP International Conference on High Performance Cutting, Procedia CIRP 14,* 2014pp. 148-152

[14]   R. Muhammad, A. Maurotto, M. Demial, A. Roy, and V.V. Silberschmidt, "Thermally enhanced ultrasonically assisted machining of Ti alloy", *CIRO J. Manuf. Sci. Technol.,* vol. 7, no. 2, pp. 159-167, 2014.
[http://dx.doi.org/10.1016/j.cirpj.2014.01.002]

[15]   M.N. Kumar, K. Subbu, and V. Krishna, *Procedia Eng.,* vol. 97, pp. 1577-1586, 2014.
[http://dx.doi.org/10.1016/j.proeng.2014.12.441]

[16]   S.B. Bhosale, R.S. Pawade, and P.K. Brahmankar, "Effect of Process Parameters on MRR, TWR and Surface Topography in ultrasonic Machining of alumina Zirconia Ceramic Composite", *Ceram. Int.,* vol. 44, no. 8, pp. pp12831-pp12836, 2014.
[http://dx.doi.org/10.1016/j.ceramint.2014.04.137]

[17]   V.V. Silberschmidta, S.M.A. Mahdyb, M.A. Goudab, A. Naseera, A. Maurottoa, and A. Roy, "Surface-roughness improvement in ultrasonically assisted turning", *2nd CIRP 2nd CIRP Conference on Surface Integrity (CSI) Procedia CIRP 13,* pp. 49-54, 2014.

[18]   D. Goswami, and S. Chakraborty, "Parametric optimization of ultrasonic machining process using gravitational search and fireworks algorithms", *Ain Shams Engineering,* vol. 6, pp. 315-331, 2015.
[http://dx.doi.org/10.1016/j.asej.2014.10.009]

[19]   S. Agarwal, "On the mechanism and mechanics of material removal in ultrasonic machining", *Int. J. Mach. Tools Manuf.,* vol. 96, pp. 1-14, 2015.
[http://dx.doi.org/10.1016/j.ijmachtools.2015.05.006]

[20]   F.D. Ning, W.L. Cong, Z.J. Pei, and C. Treadwell, "Rotary ultrasonic machining of CFRP: A comparison with grinding", *Ultrasonics,* vol. 66, pp. 125-132, 2016.
[http://dx.doi.org/10.1016/j.ultras.2015.11.002] [PMID: 26614168]

[21]   J. Wang, P. Feng, J. Zheng, and J. Zhang, "Improving hole quality in rotary ultrasonic machining of ceramic matrix composites using a compound step-taper drill", *Ceram. Int.,* vol. 42, no. 12, pp. pp13387-pp13394, 2016.
[http://dx.doi.org/10.1016/j.ceramint.2016.05.095]

[22]   S. Aminiet, H. Nouri Hosseinabadi, and S.A. Sajjady, "Experimental study on effect of micro textured surfaces generated by ultrasonic vibration assisted face turning on friction and wear performance", *Appl. Surf. Sci.,* vol. 390, pp. 633-648, 2016.
[http://dx.doi.org/10.1016/j.apsusc.2016.07.064]

[23]   R.P. Singh, and R. Singhal, "Rotary Ultrasonic Machining: a review", *Mater. Manuf. Process.,* vol. 31, no. 14, pp. 1795-1824, 2016.
[http://dx.doi.org/10.1080/10426914.2016.1140188]

[24]   Z. Li, Y. Songmei, and Z. Chonga, "Research on the Rotary Ultrasonic Facing Milling of Ceramic Matrix Composites", *9th International Conference on Digital Enterprise Technology - DET 2016 – "Intelligent Manufacturing in the Knowledge Economy Era, Procedia CIRP,* vol. 56, 2016pp. 428-433
[http://dx.doi.org/10.1016/j.procir.2016.10.077]

[25]   F. Ning, and H. Wang, "Rotary ultrasnic surface machining of CFRP composites:a comparison with conventional surface grinding", *Procedia manufacturing,* vol. 10, pp. 557-567, 2017.

[26]   P. Feng, J. Zheng, J. Zhang, and J. Wang, "Drilling induced tearing defects in rotary ultrasonic machining of c/sic composites", *Ceram. Int.,* vol. 43, no. 1, pp. 791-799, 2017.
[http://dx.doi.org/10.1016/j.ceramint.2016.10.010]

[27]   J. Wang, "K.Shimada1, M. Mizutani and T. Kuriyagawa,"Tool wear mechanism and its relation to material removal in ultrasonic machining", *Wear,* vol. 394-395, pp. 96-108, 2018.
[http://dx.doi.org/10.1016/j.wear.2017.10.010]

[28]   J. Wang, "K. Shimada1, M. Mizutani and T. Kuriyagawa, "Effects of abrasive material and particle shape on machining performance in micro ultrasonic machining", *Precis. Eng.,* vol. 51, pp. 373-387, 2018.
[http://dx.doi.org/10.1016/j.precisioneng.2017.09.008]

[29]   J. Wang, J. Zhang, P. Feng, and P. Guo, "Damage formation and suppression in rotary ultrasonic machining of hard and brittle materials: a critical review", *Ceram. Int.,* vol. 44, no. 2, pp. pp1227-pp1239, 2018.
[http://dx.doi.org/10.1016/j.ceramint.2017.10.050]

[30]   J. Wang, J. Zhang, P. Feng, and P. Guo, "Experimental and theoretical investigation on critical cutting force in rotary ultrasonic drilling of brittle materials and composites", *Int. J. Mech. Sci.,* vol. 135, pp. 555-564, 2018.
[http://dx.doi.org/10.1016/j.ijmecsci.2017.11.042]

[31]   D. Popli, and M. Gupta, "A Chipping reduction approach in Rotary Ultrasonic Machining of Advance Ceramic", *Proceedings,* vol. 5, pp. 6329-6338, 2018.

[32]   N. Mikhailova, P.Y. Onawumib, G. Volkova, I. Smirnova, M. Broseghinic, A. Roy, and V.V. Yu Petrova, "Silberschmidt,"Ultrasonically assisted drilling in marble", *J. Sound Vibrat.,* vol. 460, pp. 1-29, 2019.
[http://dx.doi.org/10.1016/j.jsv.2019.114880]

[33]   Y. Wang, "k. shimadel, M. Mizutani and T. Kuriyagawa,"Study on key factors influencing the surface generation in rotary ultrasonic grinding for hard and brittle materials", *J. Manuf. Process.,* vol. 38, pp. 549-555, 2019.
[http://dx.doi.org/10.1016/j.jmapro.2019.01.046]

[34]   H. Wang, Y. Hua, and A. R. Burks, "Rotary ultrasonic surface machining of CFRP composites effects of horizontal ultrasonic vibration", *Procedia manufacturing,* vol. 34, pp. 399-407, 2019.

[35]   H. Wang, F. Ning, Y. Li, Y. Hu, and W. Cong, "Scratching-induced surface characteristics and

material removal mechanisms in rotary ultrasonic surface machining of CFRP", *Ultrasonics,* vol. 97, pp. 19-28, 2019.
[http://dx.doi.org/10.1016/j.ultras.2019.04.004] [PMID: 31030058]

[36]   J. Wang, K. Shimada, M. Mizutani, and T. Kuriyagawa, "Effects of abrasive material and particle shape on machining performance in micro ultrasonic machining", *Precis. Eng.,* vol. 51, pp. 373-387, 2018.
[http://dx.doi.org/10.1016/j.precisioneng.2017.09.008]

[37]   X. Chen, H. Wang, Y. Hu, D. Zhang, W. Cong, and A.R. Burks, ""Rotary Ultrasonic Machining of CFRP Composites: Effects of Machining Variables on Workpiece Delamination," in Volume 2: Processes", *Materials (Basel),* 2019.
[http://dx.doi.org/10.1115/MSEC2019-3019]

[38]   R.P. Singh, and S. Singhal, "Investigation of machining characteristics in rotary ultrasonic machining of alumina ceramic", *Mater. Manuf. Process.,* vol. 32, no. 3, pp. 309-326, 2017.
[http://dx.doi.org/10.1080/10426914.2016.1176190]

[39]   H. Zhou, J. Zhang, P. Feng, D. Yu, and W. Cai, "An output amplitude model of a giant magnetostrictive rotary ultrasonic machining system considering load effect", *Precis. Eng.,* vol. 60, pp. 340-347, 2019.
[http://dx.doi.org/10.1016/j.precisioneng.2019.07.005]

[40]   N. Ahmed, B.M. Abdo, S. Darwish, K. Moiduddin, S. Pervaiz, A.M. Alahmari, and M. Naveed, "Electron beam melting of titanium alloy and surface finish improvement through rotary ultrasonic machining", *Int. J. Adv. Manuf. Technol.,* vol. 92, no. 9-12, pp. 3349-3361, 2017.
[http://dx.doi.org/10.1007/s00170-017-0365-3]

CHAPTER 2

Tool Electrode Material and Tool Fabrication Techniques for Electrical Discharge Machining Process

Arminder Singh Walia[1,*], Vineet Srivastava[1], Vivek Jain[1] and Amit Handa[2]

[1] *Mechanical Engineering Department, Thapar Institute of Engineering & Technology, Patiala-147004, Punjab, India*

[2] *Mechanical Engineering Department, I.K. Gujral Punjab technical University, Kapurthala-144603, Punjab, India*

Abstract: Electrical discharge machining is an advanced machining process used in the machining of hard materials and in die manufacturing industry. The main cost component in electrical discharge machining is the cost of the tool electrode. The conventionally used electrode material like copper and graphite has very low wear resistance, hence, wear out fast. Continuous efforts are being made to find electrode materials which have optimum value of electrical conductivity, thermal conductivity and wear resistance. Composites of various combinations of material have been tried in search of alternative tool materials. The current study presents a review of different materials utilized for the manufacturing of composite electrodes for electrical discharge machining and their fabrication techniques. It has been observed that among the different electrode fabrication techniques available, powder metallurgy has shown very promising results.

Keywords: Composite, Electrical discharge machining, Electrode wear rate, Material removal rate, Powder metallurgy, Rapid prototyping.

INTRODUCTION

In 21st century, products are started to be made from the most durable and consequently, the most difficult to machine materials. The continued development of modern structures needs materials able of combating increasing working temperatures, mechanical loads, mechanical distortions and chemical deterioration [1]. It is the need of industry to generate complex shapes of components with very tight tolerances and fine surface finish.

---

[*] **Corresponding author Arminder Singh Walia:** Mechanical Engineering Department, Thapar Institute of Engineering & Technology, Patiala-147004, Punjab, India; E-mail: arminderwalia@gmail.com

Suneev Anil Bansal (Ed.)
All rights reserved-© 2020 Bentham Science Publishers

To meet these requirements, advanced machining processes like ultrasonic machining (USM), abrasive jet machining (AJM), electrochemical machining (ECM), electron beam machining (EBM), laser beam machining (LBM), electrical discharge machining (EDM) have been developed. These methods have very key role in the tool, die, aircraft, automobile and mould making industries [2].

Abrasive water jet machining has higher material removal rate, no thermal stresses are produced on the machined surface and is also very cost friendly process. However, this process is unable to cut the thick parts and complex profiles. The quality of surface produced by this process is also poor [3]. Chemical machining is good for the machining of thin parts and requires very less equipment. In this process no residual stresses are produced. However, handling and disposal of used chemicals enhance the machining cost of this process [4]. The parts produced by laser machining suffer the heat effect from laser which causes the brittleness of machined surface and leads to surface cracks [5]. In ultrasonic machining the quality of machined surface is better than most of other modern machining process. Though, lesser material removal rate and cost incompetence are the drawback of this process [6, 7]. Electro chemical machining is a very precise process for producing complex shapes, though, the lower material removal rate and handling of chemical makes it unfavorable in some cases.

Thermoelectric energy is very useful in economic machining of low machinability materials and jobs with complex geometries. Electrical discharge machining is one of the many modern machining processes in which the material is removed by the thermal principles [2]. Out of the various advanced machining processes, EDM has been widely used to produce surgical components, dies and moulds and finishing parts for automotive and aerospace industry. Wire EDM and die sinking EDM are often used electric discharge machining techniques.

PHYSICS OF EDM

The die sinking EDM process is shown in Fig. (**1**). The EDM process can be divided into following phases:

- Cold emission of electrons and dielectric breakdown
- Spark formation, melting and evaporation
- Collapse of plasma channel and flushing

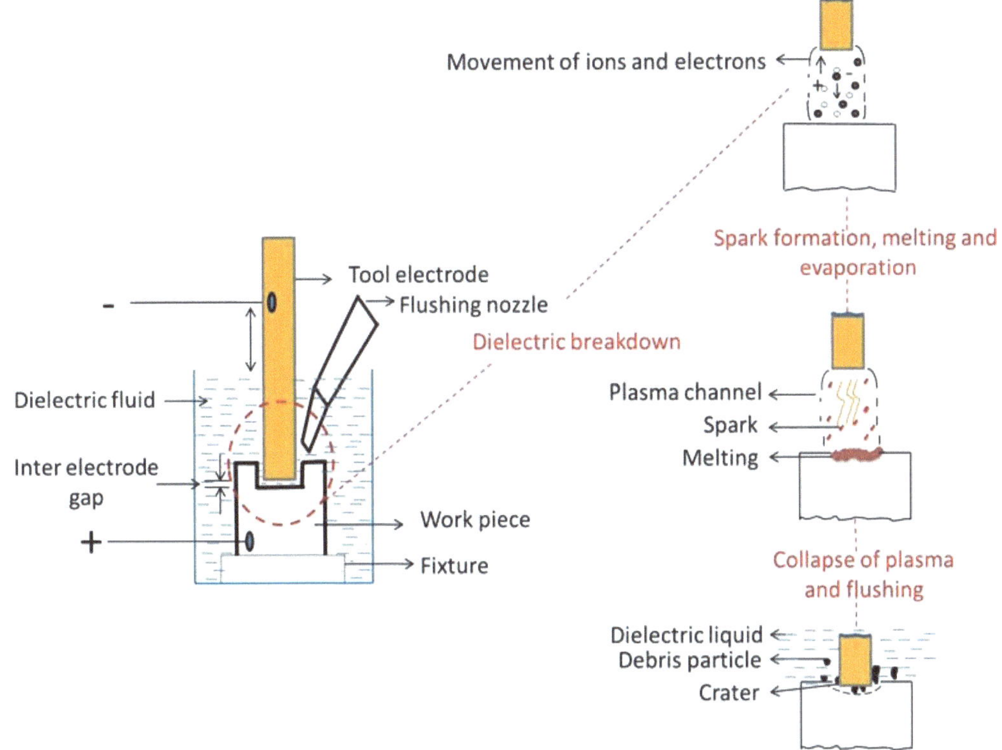

Fig. (1). Die sinking EDM process [8].

Cold Emission of Electrons and Dielectric Breakdown

A potential difference is supplied between electrically conductive tool electrode and work electrode. This potential difference creates an electric field between the two electrodes. It depends on the inter electrode gap and dielectric constant of the dielectric fluid present in between the two electrodes. The electrically conductive materials have enough number of free electrons. As the electric field is established, due to presence of electric field, the free electrons will start to get plucked from the tool (negative). The electron will be discharged in large number. This is called "Cold emission of electrons". In the presence of electric field, the electrons will accelerate towards the work as the tool is negative and workpiece is positive. The gap between these two electrodes is filled with dielectric medium. Due to presence of dielectric, no free path will be available for the electrons, hence, will face hindrance. The moving electrons will collide with the dielectric particles. The energy of the electrons will ionize the dielectric molecules. This will result in more ions as well as more electrons.

Spark Formation, Melting and Evaporation

The electrons produced in primary collision will accelerate again. It will result in secondary and tertiary collisions and so on. This will cause the release of a large number of ions and electrons in the gap between tool and workpiece. It will cause the formation of "plasma channel" which is state of matter contains electrons and ions. Once the plasma channel is formed, the resistance to the flow of electrons will decrease. The electrons will start to move from tool to workpiece at a very rapid pace and will strike the workpiece. At this time a spark can be seen. Due to this reason the EDM process is also called spark erosion process. The workpiece will continuously get strike by the electrons. The moving electrons have very high kinetic energy. On strike, this kinetic energy will get converted into heat energy. This will result in high temperature at the workpiece in the range of 8000-20000 °C [9]. The high temperature will cause localized melting and vaporization of material.

Collapse of Plasma Channel and Flushing

As soon as the supply of pulsating direct current is stopped, a pressure or shock wave will be created. This shock wave will break plasma channel thus allowing the circulating dielectric fluid to implore the plasma channel and flush the molten material from the electrode surfaces in the form of microscopic debris. A crater will be formed at the site. The repeated electrical discharges of very high temperature will remove the material and the desired cavity of tool shape will be formed.

THEORIES OF METAL REMOVAL

Depending on the experimental observations, three different kinds of material removal mechanism have been suggested [10]:

High Pressure Theory

According to this theory, as the electrodynamic waves suddenly gets stopped, enormous impulsive force is generated at the electrode. This force is responsible for the erosion of electrodes. The pressure in the discharge column is reported to be lying in the range of 10-100 kgf/cm^2. It is found that the time period for which pressure acts is more than discharge periods. It was concluded that at small discharges, the discharge pressure alone would not be enough to erode the material from electrode. However, unification of some other factors, such as heat, the pressure might remove molten metal from the electrode surface.

Static Field Theory

Coulomb's law state that there is presence of electrostatic force between two charged electrodes. The force will result in stress on the electrodes. If the gap between the electrodes is small, this force may exceed the ultimate stress limit of electrode material and result in tensile failure. In case of discharge durations lesser than 2 µs, the forces involved in tensile rupture erosion generate due to very high current densities underneath the surface of the anode create a strong electric field gradient. As the force exceeds the tensile strength of material, a tensile fracture follows thus eroding one or more particles from the electrode. But in case of discharge duration more than a few microseconds, it is observed that tensile rupture theory is not able to describe the removal of material from electrode.

High Temperature Theory

This theory suggests that as the high energy electron strikes the electrode surface, the temperature raises, particularly in case of materials with low thermal conductivity. The high temperature generated at the electrode surface immediately melts and vaporizes the material thus forming a crater. The ratio between the energy consumed at anode and total discharge energy are linked to several parameters like inter electrode gap, cathode and work function of both electrodes *etc.* It is also found that the high temperature is not produced by the striking of electrons alone. The joule heating effect generated due to high density current is also a contributing factor.

TOOL ELECTRODE MATERIALS

In EDM process, during machining, along with material removal, the wear of tool electrode is an unavoidable phenomenon. In industries, like mould and die manufacturing, the design, development and fabrication of tool takes 25-45% of the tool room lead time [11]. In majority of EDM operations, the tool cost is more than 70% of the total operation cost [12]. This signifies the need of suitable material selection for the tool. There is a perception in metal working to use right tool for the job. Choosing the correct tool material for a certain machining operation is the beginning step for generating the most productive process plan for manufacturing a part. The selection of cutting tool material depends on the work material and the operation to be performed. Often, there are possible choices of tool materials that will produce the parts successfully but not cost-effectively [13].

In conventional machining, the selection of cutting tool material depends on the work material and the operation to be performed. However, in case of EDM, the most important requirement of a tool is its electrical conductivity. The higher electrical conductivity of tool material will help to easy emission of electron during cold emission. Another requirement of EDM tool is that it should have high thermal conductivity so as to dissipate the heat. This will be very helpful to reduce the surface temperature of tool to avoid melting and evaporation of tool material. The higher density of tool material is also an advantage as for the same available heat and same amount of weight loss there will be lesser volumetric material removal from the tool, hence, lesser dimensional losses. The higher melting point of tool material also results in low tool wear as there is lesser material melting for the same amount of available heat. Apart from all the above-mentioned requirements the easy manufacturability of the tool material is also very important as it constitutes a major part of tool cost.

The most commonly used tool materials are metallic (copper, brass, tungsten, aluminium *etc.*), non-metallic (graphite) and composites (copper tungsten, copper silicon carbide, copper boron carbide, copper graphite) *etc.* Copper is the most widely used tool material owing to its high electrical and thermal conductivity. Though, in machining of difficult to machine materials, it has been seen that the copper tool erodes faster [14]. Low melting point of copper enhances wear rate of electrode that leads to variation in shape of tool. In EDM, workpiece is replica of tool profile. Accuracy of the machined cavity depends on the shape stability of the tool. Therefore, any change in the shape of tool may obtain a product that is different in shape and size than the desired one [15]. To decrease electrode wear, one of the approaches proposed by the past researchers is the fabrication of modified tool having improved mechanical, electrical and thermal properties [16 - 18]. The strength of copper can be increased by alloying materials like Cr and Zr but these elements lose these properties by dissolving in copper at high temperature [19]. To avoid this problem, the copper matrix composites are made by the addition of oxide, carbide and the addition of ceramic particles in the copper [20 - 22]. When high electrical and thermal conductivity along with high wear resistance is required Cu matrix composites are good choice [23]. Materials with high melting point and wear resistance characteristics, with good electrical and thermal conductivity are utilized to decrease the electrode wear of copper-based electrodes. Addition of ceramic reinforcements like carbides, improves the characteristics such as wear resistance and high temperature stability. The inclusion of materials such as graphite to copper matrix decreases the strength of the composites. The strength of the composites can be enhanced by the addition of ceramic reinforcement materials like silicon carbide (SiC). It increases the strength of composites, but, the electrical resistivity also increases. The composite made from SiC and Cu has the better ductility and toughness of Cu, as well as high strength of SiC reinforcements [24]. The addition of titanium Carbide (TiC) to copper improves its strength but it also slightly reduces the electrical conductivity of copper. TiC is hard material and it has very high melting temperature along with excellent thermal shock and abrasion resistance properties

[17, 25]. Often, there are possible choices of tool materials that will produce the parts successfully but not cost-effectively.

The electrode material must have high melting temperature and low electrical resistivity. The success of electrode material depends primarily upon the material being machined and on the distinct cutting application [26]. Commonly used materials for the EDM tool electrode are copper, brass, graphite, tungsten, silver and steel. High tool wear is leading concern of these types of electrodes. Continuous efforts are being carried on for developing a tool material for EDM process having low electrical and thermal resistance, high wear resistance and easy availability and fabricability [27]. Current trends in research of ceramic materials are shifting from single material to multi material electrodes. Tool properties like wear resistance, fracture toughness, hardness and thermal shock resistance have been improved significantly by the addition of one or more materials into the base material to form composite tool materials [28].

Ozgedik and Cogun [12] investigated the edge and front wear of tool during EDM of 1040 steel with copper electrode. An increase in inner and outer end wear of tool was observed with the rise in discharge current. Increase in TWR with increment in pulse duration was also recorded. In static flushing there was inappropriate flushing of the molten metal resulting in lowest material removal rate. Injection flushing produced the best surface quality.

Geric [13] made an investigation on the addition of cutting tool materials with alumina matrix. Their composition, strengthening and toughening mechanism was also examined. Mechanical properties of Al_2O_3 matrix were enhanced with the use of hard particles or whiskers. Decrease in brittleness was noticed with the addition of carbide, oxide and nitride in alumina matrix. The addition of TiB to alumina enhanced the wear resistance of the composite. The addition of yttria stabilized zirconia increase the wear resistance, however, microhardness of the composite decreased.

Srivastava and Pandey [18] tested the working of sintered Cu-TiC and conventional copper tool electrode in ultrasonic assisted cryogenically cooled EDM. EWR increased with the rise in current, however, it was lower for cermet tooltip in comparison with copper tooltip. With the increase of duty cycle there was more spark energy eventually resulted into higher EWR. Cermet electrode tip decreased out of roundness in comparison to copper tooltip. Increased MRR and surface roughness was also recorded with cermet tooltip. The surface crack density and crack width were found to be lesser on the surface machined with Cu-TiC tooltip.

Khanra *et al.* [27] fabricated ZrB_2-Cu composite using sintering process to attain an optimal value of thermal and electrical conductivity and wear resistance of tool. The composite was made by the addition of different percentages of copper. The testing was performed on mild steel for the different parameters during EDM. The composite with 40% copper resulted in higher MRR and acceptable tool wear rate (TWR) in comparison to copper tool. Though, the conventional copper tool exhibited lesser diameteral overcut and surface roughness. The deviation of MRR with variation of copper percentage is shown in Fig. (2).

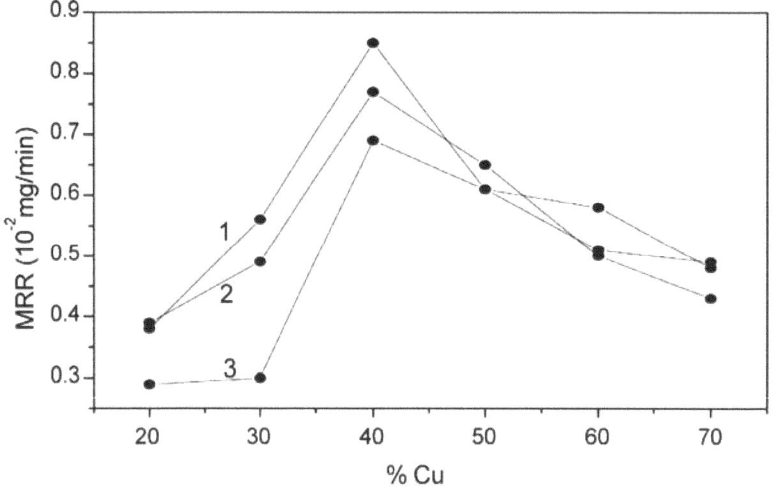

Fig. (2). Deviation of MRR with copper content [27]

Mohari *et al.* [29] used copper, aluminum, titanium and tungsten carbide powders for the fabrication of tool electrode for EDM process. The compaction pressure used for the fabrication of pellet was 343 MPa. Green, sintered and solid copper electrodes were used. Carbon steel and aluminum were used as work material. As compared to raw material, lower conductivity of green compact and sintered electrode was recorded. The wear rate of sintered and green electrode was five and ninety times more than that of solid copper electrode (0.32 μm/min).

Kumar *et al.* [30] used positive polarity Cu-Cr tool electrode to machine hastelloy with EDM. Highest MRR was achieved at 60V voltage and 18A current. Thereafter more rise in voltage and current resulted in deposition of material on work piece material from powder metallurgy tool electrode resulted in decreased material removal rate. Rise in current resulted in rise in TWR due to availability of higher energy at higher current. The best results for surface roughness were obtained at voltage 60V and current 14A because above these values of current and voltage TWR increased.

Janmanee and Muttamara [31] studied copper-tungsten (solid), copper-graphite and graphite as tool material for the EDM of tungsten carbide-cobalt (90WC-10Co) workpiece. Influence of some important process criterions like duty cycle, open voltage, electrode polarity and current were analyzed. The results showed better performance of graphite tool in terms of MRR (Fig. **3a**), whereas, the EWR and surface roughness was least for the copper-tungsten tool (Figs. **3b** , **3c**).

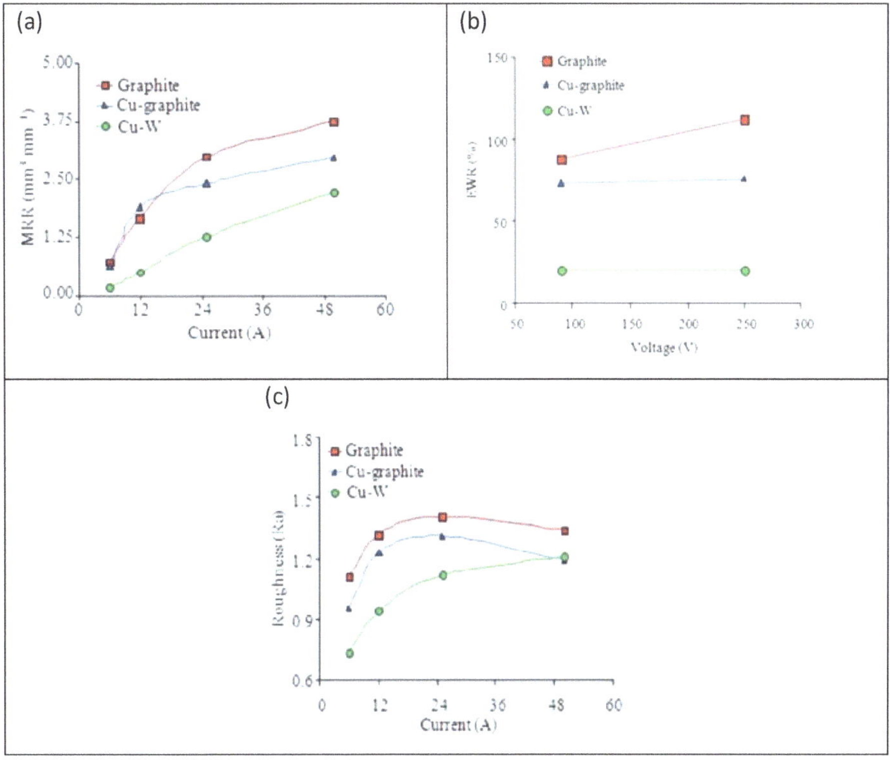

Fig. (3). Effect of discharge current on **(a)** MRR **(b)** EWR **(c)** Surface roughness [31].

Ndaliman and Khan [32] described the properties of Cu-TaC (two compositions with TaC 30% and 55% wt.) tool electrodes fabricated by powder metallurgy method. Density, electrical and thermal conductivity of the sintered and green compacted electrodes was measured and compared. The electrical conductivity, thermal conductivity and density for the green pellet was measured in the range of 94.96-189.92 $\Omega^{-1}m^{-1}$, 29.70-33.20 W/m K and 6.13-9.80 g/cm^3 respectively which were more than for the sintered pellets. The most conductive green pellet had 30% TaC.

Li *et al.* [33] reported the EDM of Inconel 718 with solid copper electrode and Cu-SiC electrode fabricated by electro-deposition. The Cu-SiC electrode enhanced the material removal efficiency by 15.12% as related with the conventional Cu electrode at 7.5 µs pulse duration and by 16.3% at 24.5 µs pulse duration. At peak current 14.2 A there was reduction in R_a by 17.1% and at 25.6 A the reduction in R_a was 16.5%. High toughness value of Inconel 718 resulted in absence of microcracks in the machined cavity. Electrode wear in Cu-SiC composite was also low due to fine microstructure and higher melting temperature. The performance of these electrodes in terms of MRR, EWR and surface roughness is given in Fig. (**4**).

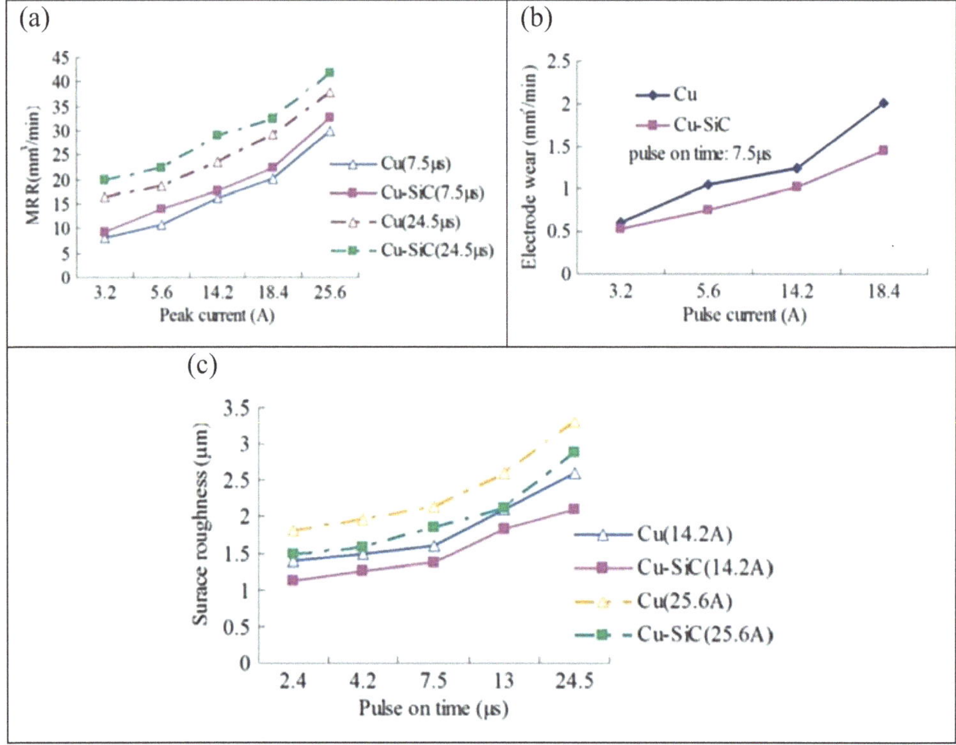

Fig. (**4**). (**a**) Variation in performance of different electrodes (**a**) MRR (**b**) EWR (**c**) Surface roughness [33]

Hussain *et al.* [34] developed a Cu-Al$_2$O$_3$ based composite using powder metallurgy technique. The reason of using Al$_2$O$_3$ was its high wear and temperature resistance. Four tools with percentage of Al$_2$O$_3$ varied at 0, 1, 3 and 5% were used during experiments. The maximum mass density and hardness of the samples was obtained with composition of 5% Al$_2$O$_3$ (Figs. **5a** and **b**). It was revealed that the wear resistance of composite was dependent on mass density and hardness and was recommended to be used as EDM electrode. The tool with 5% Al$_2$O$_3$ showed the least wear rate (Fig. **5c**).

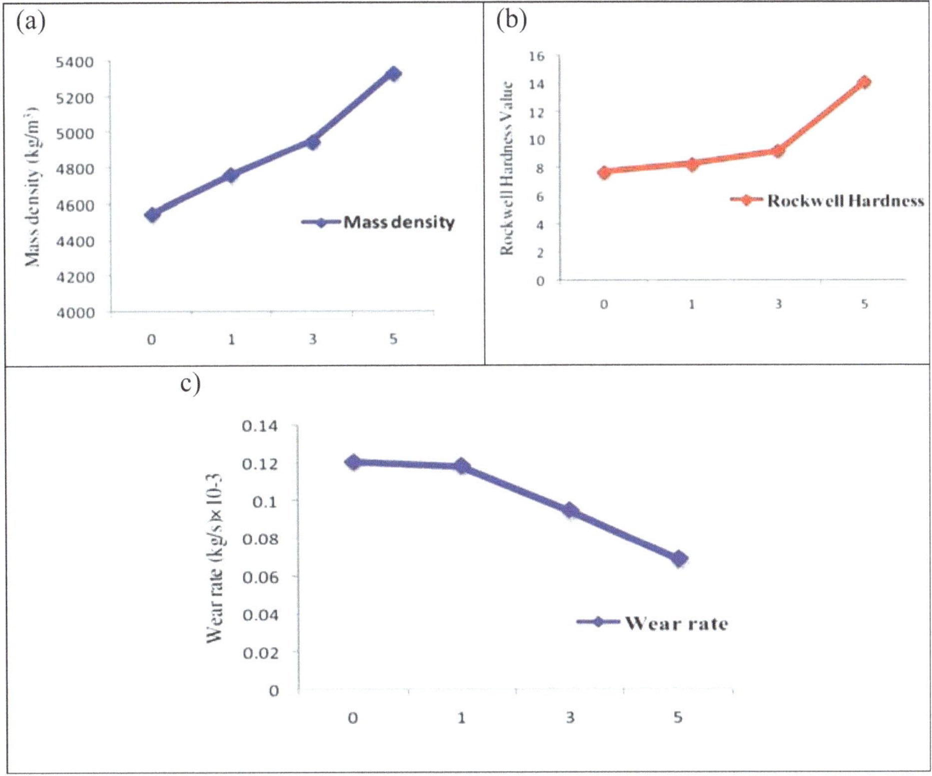

Fig. (5). Variation of **(a)** mass density **(b)** hardness **(c)** wear rate with, percentage of Al_2O_3 [34].

Kumar *et al.* [35] carried out experiments using copper-titanium diboride tool during EDM of monel 400™ alloy. Effects of percentage of titanium diboride (8-20%), pulse current (2-10 A), pulse duration (10-50 μs) and dielectric pressure (0.4-1.2 MPa) on MRR and EWR were studied. The addition of titanium diboride in the tool considerably decreased EWR and marginally improved MRR. Depending on results, an optimum set of process variables (titanium diboride 16%, pulse current 6 A, pulse on 35 μs and dielectric pressure 1 MPa) was suggested.

Liew *et al.* [36] reported the performance of carbon nanofiber (CNF) reinforced Cu tool (with CNF percentage varied at 0, 0.5, 1, 2 and 4%) during the EDM of reaction bonded SiC (RB-SiC). Among all the examined electrodes Cu-CNF (1%) showed the maximum MRR (0.5502 mm³/min) and least surface finish (Figs. **6a** and **b**). The EWR increased with the percentage of CNF (Fig. **6c**) due to bulk removal from the tool as presented in Fig. (**6d**).

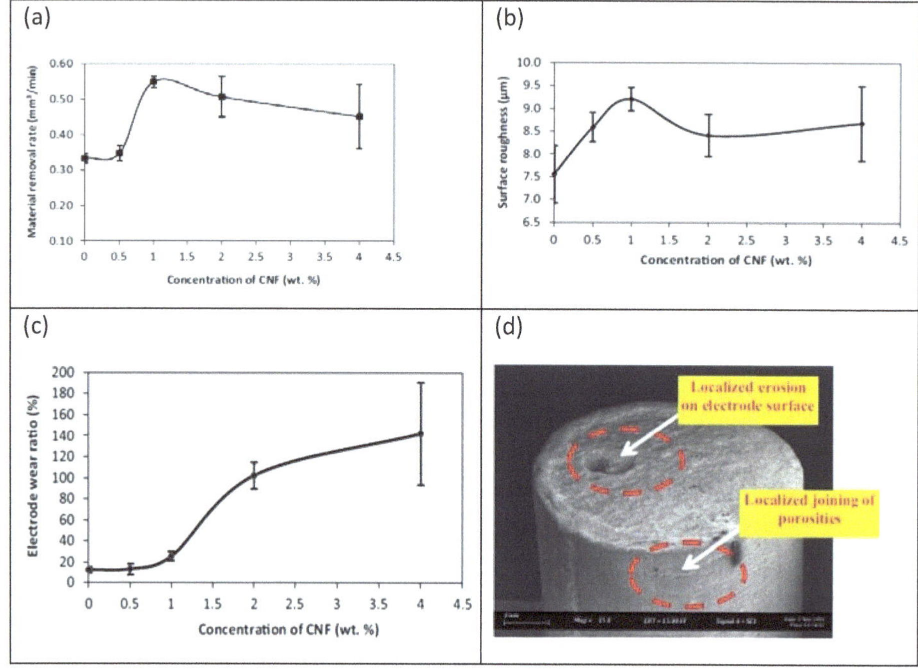

Fig. (6). Effect of CNF on **(a)** MRR **(b)** SR **(c)** EWR **(d)** Erosion of tool electrode [36]

Mandal and mondal [37] fabricated EDM electrode by coating 6061Al with copper single wall carbon nanotube. Using the hybrid tool experiments were performed on AISI 304 stainless steel. The results obtained were compared with copper coated 6061Al and uncoated 6061Al tool. The results exhibited 3.13% and 14.52% increase in MRR and 21.04% and 29.6% decrease in EWR for copper-coated and copper single wall carbon nanotube coated 6061Al electrodes, respectively, in comparison to uncoated 6061Al electrode. Surface roughness of the machined cavity was decreased by 14.06% and 8.03% using copper-coated and copper single wall carbon nanotube coated 6061Al electrodes.

The major research efforts in the use of different type of tool electrode material are tabulated in Table **1**.

Table 1. Major research efforts in the use of different type of tool electrode material

Investigators	Material of Tool Electrode	Material of Work Electrode	Investigation Factors	Responses
Srivastava and Pandey [18]	Cu-TiC	M2 HSS	Gap voltage, current, pulse duration, duty cycle	Decreased out of roundness, Increased MRR and surface roughness was also recorded when cermet tooltip was used.

(Table 1) cont.....

Investigators	Material of Tool Electrode	Material of Work Electrode	Investigation Factors	Responses
Mohari *et al.* [29]	Cu, WC-Co, TiC-Al	Carbon steel and aluminum	Pulse duration, Machining time	The lower conductivity of green compact or sintered product resulted in higher wear rate.
Kumar *et al.* [30]	Cu90-Cr10	Hastelloy	Current, Voltage	The best result for surface roughness was obtained at 14A and voltage 60V because above these values TWR increased.
Janmanee and Muttamara [31]	Graphite, Copper-Graphite, Cu-W (solid)	Tungsten Carbide-Cobalt (WC90-Co10)	On/off time, current, open voltage	Better performance of graphite and copper-graphite powder metallurgy in terms of MRR and TWR.
Li *et al.* [33]	Solid copper electrode, Cu-SiC	Inconel 718 alloy steel	Pulse duration, pulse interval, voltage, current	MRE with Cu-SiC electrode increased by 15.12% as related to copper electrode.
Hussain *et al.* [34]	Cu-Al$_2$O$_3$	High speed steel	Percentage of Al$_2$O$_3$ in tool	The tool with 5% Al$_2$O$_3$ showed the least wear rate.
Mandal and mondal [37]	Uncoated 6061Al, Cu single wall carbon nanotube coated 6061Al, Cu coated 6061Al	AISI 304 stainless steel	Tool composition	The results exhibited 3.13% and 14.52% increase in MRR and 21.04% and 29.6% decrease in EWR for copper-coated and copper single wall carbon nanotube coated 6061Al electrodes, respectively.

FABRICATION OF ELECTRODES USING ALTERNATE APPROACHES

The physical properties of electrode influence the machining performance. High electrical conductivity, high specific heat, good machinability, high erosion resistance *etc.* are the properties which an ideal electrode must possess. A single material cannot fulfill all these requirements. Traditional machining methods are mostly used for fabricating the electrodes for EDM [38]. Complexity of the geometry and the precision required commonly determines the manufacturing cost of the electrodes. The uniform dimension electrodes are cheaper and easy to produce but a complex shape electrode made by traditional machining process may cost 100 times than a simple shape electrode [39]. For the fabrication of complicate three-dimensional shape electrodes, different methods like conventional machining, electro-forming, metal spraying, casting, rapid prototyping or powder metallurgy are used [17, 40]. These techniques can also be helpful to reduce the lead time. A review of research attempts to fabricate the tool for EDM is presented below.

Samuel and Philip [16] carried out experimental work and observed that with P/M method tool electrodes can be fabricated with ease and its properties can be controlled rather easily, consequently gave it an edge over other electrode fabrication methods. The effect of compaction pressure on relative density and on green strength is shown in Figs. (**7a** and **b**). At low sintering temperature and compacting pressure gap contamination was more and higher electrode wear was observed. Low metal removal at low current and high frequency was noticed due to surface properties of powder metallurgy electrodes. Low compact pressure and sintering temperature resulted in poor mechanical strength and high tool wear.

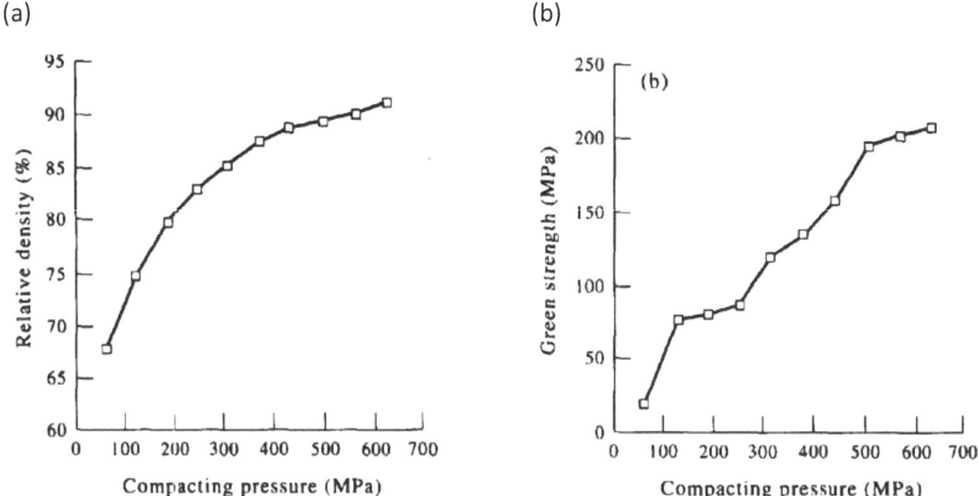

Fig. (7). Influence of compaction pressure on (**a**) Relative density (**b**) Green strength [16].

Samuel and Philip [39] during the study used copper (Cu) powder with 99.7% purity as tool material and fabricated the tool using powder metallurgy (P/M) process. Increase in conductivity was found with the rise in sintering temperature and compacting pressure. The electrode compacted at low pressure had low mechanical strength. High sintering temperature increased the bond strength and decreases erosion. Thus, the increase in compacting pressure and sintering temperature resulted in change in electrical, thermal, mechanical and microstructural properties resulting better performance.

Dimla *et al.* [40] applied stereolithography and direct metal laser sintering for the coating of models to fabricate the tool for EDM process. The quantity of copper deposited on models by both methods created problems. The used processes were unable to deposit even layer of copper on the models and it varied from 50-1000 µm. These methods were not able to deposit required amount of copper, especially in the inner parts of the electrodes. The thickness of coating decreased from the outer surface to the inner walls and the tool electrodes were not found suitable for EDM. The variation in copper thickness inside vertical wall of electrode is shown in Fig. (**8**).

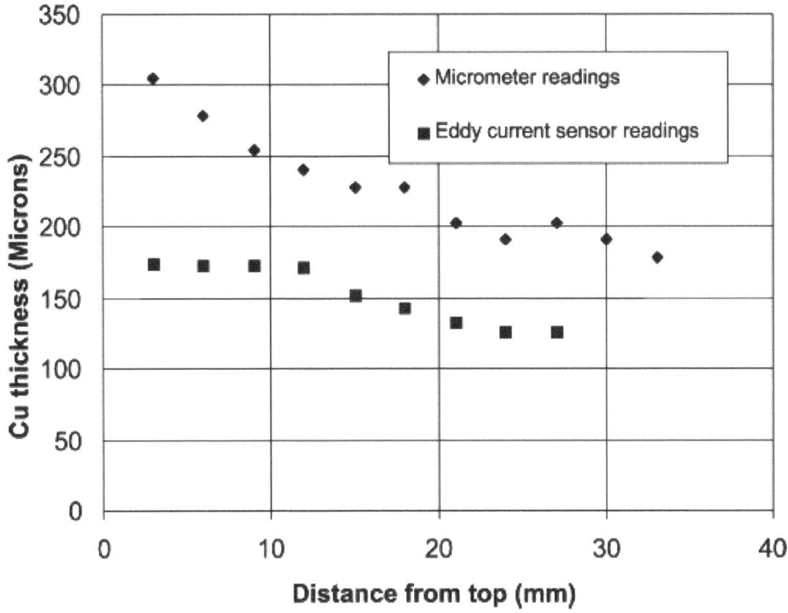

Fig. (8). Measurement of copper deposition on inside vertical wall of electrode [40].

Durr *et al.* [41] introduced the concept of direct metal laser sintering for the fabrication of tool electrode for EDM process. The electrodes contained bronze, nickel and copper phosphide powders, were built layer by layer. Process optimization made it clear that the porosity of the sintered electrodes directly controlled by the laser speed, laser power and sintering strategy. The electrodes were also infiltrated with silver containing brazing metal. These electrodes achieved MRR approximately 12.5 mm^3/min but it exhibited greater EWR and surface roughness as compared with conventional electrodes.

Norasetthekul *et al.* [42] synthesized tool electrodes by mechanical pressing of polymer coated ZrB_2 powders, followed by infiltration of copper into the green body in a high-temperature furnace. It was noticed that higher the compaction pressure, the lesser the voids between the particles, thus reducing the amount of copper required to infill the voids. However, it was found that the copper tends to fill the voids incompletely. It was observed that the ZrB_2-Cu electrode had the highest thermal wear resistance to the plasma spark at the condition of extremely high energy flux. Also, the tool was able to cut the steel work piece faster than graphite and copper tools.

Li *et al.* [43] revealed the effect of addition of TiC the properties of sintered Cu based tools. The pellets were fabricated by P/M of copper and copper-tungsten (Cu-W) powders by varying the TiC from 5% to 45% followed by liquid phase sintering. Addition of TiC improved the surface finish and suggested to be used for finishing purpose. TWR was also found to be lesser for all these electrodes in comparison to commercially available electrodes.

Zhao *et al.* [44] reported the use of selective laser sintering (SLS) as useful method to fabricate EDM metal prototype. The density and other facets of the mechanical performance of an SLS metal prototype were improved by the metal infiltration. The metal prototype was dipped into a liquid infiltrant (copper used commonly) that rose into the open pores by the mode of capillary action. The results showed that the wear rate of the electrode was comparable to that of conventional electrodes. The surface roughness obtained was within acceptable range for the similar machine settings.

Blom *et al.* [45] employed spray metal deposition and electroforming procedures to produce tool electrodes. Performance of these electrodes was compared with conventional solid tool electrodes. For different working conditions MRR, TWR, machining time and surface roughness were measured. The solid tool achieved better results as compared with electroformed tool in terms of MRR and TWR. However, the tools synthesized by electroforming method were found to be cheaper than the solid tools.

Shibayama and Kunieda [46] revealed the electric discharge machining of deep slots with the electrode having very small size holes. The holes directed the dielectric fluid over the work surface. Two copper plates having micro grooves over the interface for directing the dielectric fluid were joined by diffusion bonding to form the electrode. This electrode took less processing time with more machining preciseness as compared with normally used solid tool electrode.

Mishra and Pathak [47] added carbon (0 to 10%) and titanium carbide (0 to 10%) in ZrB_2 powder and studied the systematic sintering. The mixing of carbon upto 4% resulted in densification but further increase in carbon addition retarded the densification. Addition of TiC above 5% had adverse effect on sintering of ZrB_2. The inclusion of carbon helped to hinder the grain size and composite with fine grain size and higher density was obtained.

Hsu *et al.* [48] used rapid prototyping based on electroless plating (nickel plating) and electroforming (copper) for the fabrication of electrode for EDM process.

Process of this type reduced the manufacturing duration as well as curtailed the cost of electrode fabrication. Experimental observations showed no crack on the electrode and promising results for EDM process.

Monzon *et al.* [49] tried rapid prototyping and electroforming process for the manufacturing of EDM electrodes. They developed copper electroformed shell testing methodology. The various aspects such as composition, internal stresses, electrolytic bath additives, pitting, dimensional precision and roughness were analyzed. As compared to conventional techniques favorable outcomes were recorded in terms of weight and ease of fabrication of electrode.

Egashira *et al.* [50] fabricated ultra-small tungsten tools by the combination of wire electrode discharge grinding and electrochemical machining. Holes with radius lesser than 0.5 µm and more than 1 µm in depth were drilled in brass and zinc material successfully. However, the circularity of the hole was not found good as the size of a single discharge crater was found to be greater than the hole diameter.

Meshram and Puri [51] used three different methods, rapid wire EDM, net shape casting process and investment casting process for the fabrication of EDM electrodes. Rapid wire EDM process resulted in better dimensional accuracy in comparison to other two methods. Net shape casting required lesser time for the electrode fabrication. The copper-brass electrode fabricated by this process resulted in highest MRR and EWR. Investment casting was found to have edge over the other methods to produce complex shape electrodes.

Balasubramanian and Senthilvelan [52] performed the EDM of EN8 and D3 steel with cast and powder metallurgy copper tool electrode. Higher average MRR (77.4 mm^3/min) and lower average TWR (10.99 mm^3/min) was obtained with cast electrode as compared to sintered electrode for EN-8 material. Cast electrode also showed high MRR (136.1 mm^3/min) and low TWR (3.32 mm^3/min) in EDM of D3 steel. Sintered electrode showed better surface finish for EN-8 and D3 steel with comparison to cast electrode.

Pal and Choudhry [53] manufactured copper and brass electrode for EDM process by abrasive water jet machining (AWJM). Array of holes were made on stainless steel and Ti-6Al-4V by keeping voltage, current, duty cycle and pulse duration constant. Deviations in the dimensions of the tool produced by abrasive water jet machining and dimensions of the workpiece were observed as shown in Fig. (**9**). Two reasons were held responsible for this. First was taper produced on the tool due to instinctive property of abrasive water jet machining jet which produced kerf on the fin length. Second logic was that the side walls of the blind cavities got machined because the tool was not masked on the sides.

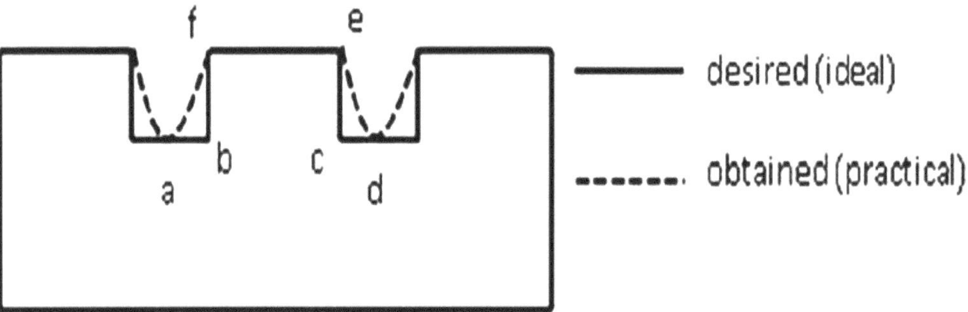

Fig.(9). Ideal and practical cut of AWJM [53].

Goyal *et al.* [54] investigated the hardness and SR of EN31 steel machined by EDM process using Cu-Mn tool made by powder metallurgy technique. Discharge current and pulse duration were chosen as machining variable. The microhardness of the surface machined with 70Cu-30Mn electrode was found to be 913 VHN which was 11.94% more than copper electrode and 3.39%more than 80Cu-20Mn electrode. However, the powder metallurgy electrodes produced surface with higher roughness due to its porous nature, though the difference was marginal.

Sahu *et al.* [55] explored the EDM of Nitinol using AlSiMg tool fabricated by SLS and compared its performance with the commercially available graphite and copper tools. Surface roughness was selected as response to examine the influence of electrode material, current, voltage, pulse duration and duty cycle. The average surface roughness achieved with SLS tool was least followed by copper tool.

Equbal *et al.* [56] fabricated EDM tool using fused deposition modelling (FDM). Acrylonitrile butadiene styrene (ABS) was used as the core material and outer shell was produced by copper coating. With FDM, it was found very difficult to maintain the sharp edges of tool. Although, the performance of tool fabricated with FDM process was comparable with solid copper tool, but it has lower dimensional accuracy. The difficulty in producing the exact dimension of tool with FDM was cited as the reason.

Saxena and Metkar [57] also developed an EDM tool using FDM. Copper coating was done on a core of ABS material fabricated by FDM. influence of current, pulse time and voltage on EWR and MRR was analyzed. Similar results for MRR were obtained with solid copper tool and tool fabricated by FDM, whereas, the later performed slightly better in terms of EWR.

The major research efforts in the fabrication techniques of EDM tool electrode are presented in Table **2**.

Table 2. Major research efforts in fabrication of EDM tool electrode.

Investigator	Method of Tool Electrode Fabrication	Material of the Tool Electrode	Remarks
Dimla *et al.* [40]	Rapid prototyping, Electroplating	Cu, Bronze	Enough Cu was not deposited on inner cavities.
Durr *et al.* [41]	Direct metal laser sintering	Bronze, nickel and copper phosphide	These electrodes achieved MRR 12.5 mm^3/min but it exhibited greater EWR and SR as compared with conventional electrodes.
Li *et al.* [43]	Rapid prototyping	Cu-TiC-W	Nickel improved densification of TiC-Cu-W as nickel showed good solubility in Cu and W.
Blom *et al.* [45]	Electroforming, Spray-metal deposition	Solid Cu, Electroformed Cu	For fabrication of large number of electrodes electroforming takes similar time as required for single electrode.
Monzon [49]	Electroforming, Rapid prototyping	Cu	Results showed favorable results in terms of weight and ease of fabrication.
Egashira *et al.* [50]	wire electrode discharge grinding and electrochemical machining	Tungsten	Holes of lesser than 0.5 μm radius and greater than 1 μm in depth were drilled in brass and zinc material successfully.
Meshram and Puri [51]	Rapid wire EDM, net shape casting process and investment casting	Al, Cu, Cu-Brass	Rapid wire EDM process resulted in better dimensional accuracy. Net shape casting required lesser time for the electrode fabrication. Investment casting has edge over the other methods to produce complex shape electrodes.
Balasubramanian and Senthilvelan [52]	Casting and sintered powder metallurgy electrode	Cu	The cast copper electrode showed higher MRR, low TWR and higher surface roughness for EN8 and D3.

(Table 2) cont.....

Investigator	Method of Tool Electrode Fabrication	Material of the Tool Electrode	Remarks
Pal and Choudhry [53]	Abrasive water jet machining	Cu and Brass	Deviations in the dimensions of the tool due to taper produced on the tool in AWJM and due to machining of side walls.
Sahu *et al.* [55]	Selective laser sintering	AlSiMg	The surface machined by AlSiMg tool electrode has lesser surface crack density.
Equbal *et al.* [56]	Fused deposition modelling	Acrylonitrile-butadiene-styrene and Copper	It was found very difficult to maintain the sharp edges of tool and has lower dimensional accuracy.

CONCLUSIONS

It has also been observed that the tool end gets flattened and deformed over a period of time. It can be avoided if the electrode end is replaced with hard conducting materials like cermets. These cermet tooltips can have the desirable properties of different combined materials. For the fabrication of tool, many methods have been used. Complexity of the geometry and the precision required commonly determines the manufacturing cost and fabrication technique to be used for the fabrication of the electrodes. The conventional methods take more processing time and the wastage of material is also more, especially during the fabrication of complex shape tool profiles. The cermet tool tips can be fabricated by casting, solid phase sintering, rapid prototyping and tooling, spray metal deposition, direct metal laser sintering, abrasive water jet machining, powder metallurgy *etc.* Some of the casting methods like investment casting are expensive and takes more time. The processes like net shape casting have lesser dimensional accuracy [51]. The electrodes produced by selective laser sintering process have very high porosity (upto 20%), higher wear and produces machined cavity with very low accuracy [41]. The electrodes produced by spray metal deposition have the problem of uneven thickness of coating and sometime the spray metal also penetrate the base material [45]. In some of processes, like stereolithography and direct metal laser sintering, uneven coating thickness was observed rendering the electrodes unsuitable for EDM process [40]. The electrodes produced by electroforming have very high surface roughness [49]. Abrasive water jet machining jet which produced kerf on the fin length due to this taper produced on the produced components/ electrodes. Another problem associated with this process is that the side walls of the blind cavities got machined because the tool was not masked on the sides Pal and Choudhry [53].

In powder metallurgy (P/M) technique, it is possible to combine the desirable properties of different materials and change the physical properties through compacting pressure and sintering temperature [40]. Thus, through P/M there is better control of the properties of the electrode possible. Ease of fabrication, flexibility of composition and efficacy in dispersing fine particles [34] are the benefits of powder metallurgy thereby making it a suitable method for electrode fabrication.

CONSENT FOR PUBLICATION

Not applicable.

CONFLICT OF INTEREST

The authors confirm that this chapter content has no conflict of interest.

ACKNOWLEDGEMENTS

The authors would like to express their sincere thanks to the editor and anonymous reviewers for their time and valuable suggestions.

REFERENCES

[1]   T.S. Sidhu, S. Prakash, and R.D. Agrawal, "Hot corrosion performance of a NiCr coated Ni-based alloy", *Scr. Mater.,* vol. 55, pp. 179-182, 2006.
        [http://dx.doi.org/10.1016/j.scriptamat.2006.03.054]

[2]   H.A.G. El-Hofy, *Advanced machining processes: nontraditional and hybrid machining processes.* McGraw Hill Professional, 2016.

[3]   A.W. Momber, I. Eusch, and R. Kovacevic, "Machining refractory ceramics with abrasive water jets", *J. Mater. Sci.,* vol. 31, pp. 6485-6493, 1996.
        [http://dx.doi.org/10.1007/BF00356252]

[4]   H.T. Ting, K.A. Abou-El-Hossein, and H.B. Chua, "Review of micromachining of ceramics by etching", *Trans. Nonferrous Met. Soc. China,* vol. 19, pp. s1-s16, 2009.
        [http://dx.doi.org/10.1016/S1003-6326(10)60237-3]

[5]   A.N. Samant, and N.B. Dahotre, "Laser machining of structural ceramics-A review", *J. Eur. Ceram. Soc.,* vol. 29, pp. 969-993, 2009.
        [http://dx.doi.org/10.1016/j.jeurceramsoc.2008.11.010]

[6]   M.N. Farooqui, and N.G. Patil, "A perspective on shaping of advanced ceramics by electro discharge machining", *Procedia Manufacturing,* vol. 20, pp. 65-72, 2018.
        [http://dx.doi.org/10.1016/j.promfg.2018.02.009]

[7]   J. Kumar, "Ultrasonic machining- A comprehensive review", *Mach. Sci. Technol.,* vol. 17, pp. 325-379, 2013.
        [http://dx.doi.org/10.1080/10910344.2013.806093]

[8]   "Walia, V. Srivastava, and V. Jain, "Impact of copper-titanium carbide tooltip on machined surface integrity during Electrical Discharge Machining of EN31 die steel", *Mater. Res. Express,* vol. 6, no. 106582, 2019.

[9]   P. Ong, C.H. Chong, M.Z. Bin Rahim, W.K. Lee, C.K. Sia, and M.A.H. Bin Ahmad, "Intelligent approach for process modelling and optimization on electrical discharge machining of polycrystalline diamond", *J. Intell. Manuf.,* pp. 1-21, 2018.

[10]   P.K. Mishra, *Nonconventional machining.* Narosa publishing house: India, 2007.

[11]   G.S. Brar, and H.S. Sandhu, "Investigation of Electric Discharge Machining by Electroformed Tool Fabricated using Additive Manufacturing", *AIMTDR,* pp. 713-717, 2016.

[12]   A. Ozgedik, and C. Cogun, "An experimental investigation of tool wear in electric discharge machining", *Int. J. Adv. Manuf. Technol.,* vol. 27, pp. 488-500, 2016.
[http://dx.doi.org/10.1007/s00170-004-2220-6]

[13]   K. Geric, "Ceramics tool materials with alumina matrix", *Mach. Des,* vol. 3, pp. 367-372, 2010.

[14]   C. Upadhyay, S. Datta, M. Masanta, and S.S. Mahapatra, "An experimental investigation emphasizing surface characteristics of electro-discharge-machined Inconel 601", *J. Braz. Soc. Mech. Sci. Eng.,* vol. 39, pp. 3051-3066, 2017.
[http://dx.doi.org/10.1007/s40430-016-0643-2]

[15]   A.S. Walia, V. Srivastava, V. Jain, and M. Garg, Modelling and Analysis of Change in Shape of sintered Cu–TiC tool tip during Electrical Discharge Machining process.*Advances in Unconventional Machining and Composites. Lecture Notes on Multidisciplinary Industrial Engineering.* Springer: Singapore, 2020, pp. 515-525.
[http://dx.doi.org/10.1007/978-981-32-9471-4_42]

[16]   M.P. Samuel, and P.K. Philip, "Power metallurgy tool electrodes for electrical discharge machining", *Int. J. Mach. Tools Manuf.,* vol. 37, pp. 1625-1633, 1997.
[http://dx.doi.org/10.1016/S0890-6955(97)00006-0]

[17]   H.M. Zaw, J.Y.H. Fuh, A.Y.C. Nee, and L. Lu, "Formation of a new EDM electrode material using sintering techniques", *J. Mater. Process. Technol.,* vol. 89, pp. 182-186, 1999.
[http://dx.doi.org/10.1016/S0924-0136(99)00054-0]

[18]   V. Srivastava, and P.M. Pandey, "Study of ultrasonic assisted cryogenically cooled EDM process using sintered (Cu–TiC) tooltip", *J. Manuf. Process.,* vol. 15, pp. 158-166, 2013.
[http://dx.doi.org/10.1016/j.jmapro.2012.12.002]

[19]   J.B. Correia, H.A. Davies, and C.M. Sellars, "Strengthening in rapidly solidified age hardened Cu–Cr and Cu–Cr–Zr alloys", *Acta Mater.,* vol. 45, pp. 177-190, 1997.
[http://dx.doi.org/10.1016/S1359-6454(96)00142-5]

[20]   A.T. Alpas, H. Hu, and J. Zhang, "Plastic deformation and damage accumulation below the worn surfaces", *Wear,* vol. 162, pp. 188-195, 1993.
[http://dx.doi.org/10.1016/0043-1648(93)90500-L]

[21]   A.K. Shukla, S.N. Murty, R.S. Kumar, and K. Mondal, "Effect of powder milling on mechanical properties of hot-pressed and hot-rolled Cu–Cr–Nb alloy", *J. Alloys Compd.,* vol. 580, pp. 427-434, 2013.
[http://dx.doi.org/10.1016/j.jallcom.2013.06.118]

[22]   A.S. Walia, V. Srivastava, and V. Jain, "Fabrication and Application of Composite Electrodes in Electrical Discharge Machining-A Review", *International Journal of Computer Applications (0975-8887),* pp. 1-5, 2017.

[23]   S. Buytoz, F. Dagdelen, S. Islak, M. Kok, D. Kir, and E. Ercan, "Effect of TiC content on microstructure and thermal properties of Cu-TiC composites prepared by powder metallurgy", *J Thermal Anal Carolim.,* vol. 117, pp. 1277-1283, 2014.
[http://dx.doi.org/10.1007/s10973-014-3900-6]

[24]   G.C. Efe, I. Altinsoy, M. Ipek, S. Zeytin, and C. Bindal, "Some properties of Cu–SiC composites produced by powder metallurgy method", *Kovove Mater,* vol. 49, p. 131, 2011.

[http://dx.doi.org/10.4149/km_2011_2_131]

[25]   K.M. Shu, and G.C. Tu, "Fabrication and characterization of Cu-SiCp composites for electrical discharge machining applications", *Mater. Manuf. Process.*, vol. 16, pp. 483-502, 2001.
[http://dx.doi.org/10.1081/AMP-100108522]

[26]   S. Gopalakannan, and T. Senthilvelan, "Effect of electrode materials on electric discharge machining of 316 L and 17-4 PH stainless steels", *J. Miner. Mater. Charact. Eng.*, vol. 11, pp. 685-690, 2012.
[http://dx.doi.org/10.4236/jmmce.2012.117053]

[27]   A.K. Khanra, B.R. Sarkar, B. Bhattacharya, L.C. Pathak, and M.M. Godkhindi, "Performance of ZrB2–Cu composite as an EDM electrode", *J. Mater. Process. Technol.*, vol. 183, pp. 122-126, 2007.
[http://dx.doi.org/10.1016/j.jmatprotec.2006.09.034]

[28]   T. Czelusniak, F.L. Amorim, C.F. Higa, and A. Lohrengel, "Development and application of new composite materials as EDM electrodes manufactured via selective laser sintering", *Int. J. Adv. Manuf. Technol.*, vol. 72, pp. 1503-1512, 2014.
[http://dx.doi.org/10.1007/s00170-014-5765-z]

[29]   N. Mohri, N. Saito, Y. Tsunekawa, and N. Kinoshita, "Metal surface modification by electrical discharge machining with composite electrode", *CIRP Ann.*, vol. 42, pp. 219-222, 1993.
[http://dx.doi.org/10.1016/S0007-8506(07)62429-9]

[30]   V. Kumar, N. Beri, A. Kumar, and P. Singh, "Some studies on electric discharge machining of hastelloy using powder metallurgy electrode", *Int. J. Adv. Eng. Technol.*, vol. 1, pp. 16-27, 2010.

[31]   P. Janmanee, and A. Muttamara, "Performance of difference electrode materials in electrical discharge machining of tungsten carbide", *Energy Research Journal,* vol. 1, pp. 87-90, 2010.
[http://dx.doi.org/10.3844/erjsp.2010.87.90]

[32]   M.B. Ndaliman, and A.A. Khan, "Development of powder metallurgy (pm) compacted cu-tac electrode for edm", *Journal of Mechanics Engineering and Automation,* vol. 1, pp. 385-391, 2011.

[33]   L. Li, Z.Y. Li, X.T. Wei, and X. Cheng, "Machining characteristics of Inconel 718 by sinking-EDM and wire-EDM", *Mater. Manuf. Process.*, vol. 30, pp. 968-973, 2015.
[http://dx.doi.org/10.1080/10426914.2014.973579]

[34]   M.Z. Hussain, U. Khan, R. Jangid, and S. Khan, "Hardness and wear analysis of Cu/Al2O3 composite for application in EDM electrode", *IOP Conf. Series Mater. Sci. Eng.*, vol. 310, no. 012044, 2018.

[35]   P.M. Kumar, K. Sivakumar, and N. Jayakumar, "Multiobjective optimization and analysis of copper–titanium diboride electrode in EDM of monel 400™ alloy", *Mater. Manuf. Process.*, vol. 33, pp. 1429-1437, 2018.
[http://dx.doi.org/10.1080/10426914.2017.1415439]

[36]   P.J. Liew, Z. Nurlishafiqa, Q. Ahsan, T. Zhou, and J. Yan, "Experimental investigation of RB-SiC using Cu–CNF composite electrodes in electrical discharge machining", *Int. J. Adv. Manuf. Technol.*, vol. 98, pp. 3019-3028, 2018.
[http://dx.doi.org/10.1007/s00170-018-2417-8]

[37]   P. Mandal, and S.C. Mondal, "Development and application of Cu-SWCNT nanocomposite–coated 6061Al electrode for EDM", *Int. J. Adv. Manuf. Technol.*, pp. 1-10, 2019.
[http://dx.doi.org/10.1007/s00170-019-03710-5]

[38]   A.S. Walia, V. Jain, and V. Srivastava, "Development and performance evaluation of sintered tool tip while EDMing of hardened steel", *Mater. Res. Express,* vol. 6, no. 086520, 2019.
[http://dx.doi.org/10.1088/2053-1591/ab1c7a]

[39]   M.P. Samuel, and P.K. Philip, "Properties of compacted, pre-sintered and fully sintered electrodes produced by powder metallurgy for Electric Discharge Machining", *Indian J. Eng. Mater. Sci.*, vol. 3, pp. 229-233, 1996.

[40]   D.E. Dimla, N. Hopkinson, and H. Rothe, "Investigation of complex rapid EDM electrodes for rapid

tooling applications", *Int. J. Adv. Manuf. Technol.,* vol. 23, pp. 249-255, 2004.
[http://dx.doi.org/10.1007/s00170-003-1709-8]

[41]   H. Durr, R. Pilz, and N.S. Eleser, "Rapid tooling of EDM electrodes by means of selective laser sintering", *Comput. Ind.,* vol. 39, pp. 35-45, 1999.
[http://dx.doi.org/10.1016/S0166-3615(98)00123-7]

[42]   S. Norasetthekul, P.T. Eubank, W.L. Bradley, B. Bozkurt, and B. Stucker, "Use of zirconium diboride-copper as an electrode in plasma applications", *J. Mater. Sci.,* vol. 34, pp. 1261-1270, 1999.
[http://dx.doi.org/10.1023/A:1004529527162]

[43]   L. Li, Y.S. Wong, J.Y.H. Fuh, and L. Lu, "EDM performance of TiC/copper-based sintered electrodes", *Mater. Des.,* vol. 22, pp. 669-678, 2001.
[http://dx.doi.org/10.1016/S0261-3069(01)00010-3]

[44]   J. Zhao, Y. Li, J. Zhang, C. Yu, and Y. Zhang, "Analysis of the wear characteristics of an EDM electrode made by selective laser sintering", *J. Mater. Process. Technol.,* vol. 138, pp. 475-478, 2003.
[http://dx.doi.org/10.1016/S0924-0136(03)00122-5]

[45]   R.J. Blom, P.K. Yarlagadda, and R.M. Iyer, "Evaluation Of Rapid Tooling For Electric Discharge Machining Using Electroforming And Spray Metal Deposition Techniques", *Proceedings 1ˢᵗ Global Congress on Manufacturing and Management,* 2004pp. 490-495 Vellore, India

[46]   T. Shibayama, and M. Kunieda, "Diffusion bonded EDM electrode with micro holes for jetting dielectric liquid", *CIRP Ann.,* vol. 55, pp. 171-174, 2006.
[http://dx.doi.org/10.1016/S0007-8506(07)60391-6]

[47]   S.K. Mishra, and L.C. Pathak, "Effect of carbon and titanium carbide on sintering behaviour of zirconium diboride", *J. Alloys Compd.,* vol. 465, pp. 547-555, 2008.
[http://dx.doi.org/10.1016/j.jallcom.2007.11.004]

[48]   C.Y. Hsu, D.Y. Chen, M.Y. Lai, and G.J. Tzou, "EDM electrode manufacturing using RP combining electroless plating with electroforming", *Int. J. Adv. Manuf. Technol.,* vol. 38, pp. 915-924, 2008.
[http://dx.doi.org/10.1007/s00170-007-1155-0]

[49]   M. Monzon, A.N. Benítez, M.D. Marrero, N. Hernandez, P. Hernandez, and J. Aisa, "Validation of electrical discharge machining electrodes made with rapid tooling technologies", *J. Mater. Process. Technol.,* vol. 196, pp. 109-114, 2008.
[http://dx.doi.org/10.1016/j.jmatprotec.2007.05.025]

[50]   K. Egashira, Y. Morita, and Y. Hattori, "Electrical discharge machining of submicron holes using ultrasmall-diameter electrodes", *Precis. Eng.,* vol. 34, pp. 139-144, 2010.
[http://dx.doi.org/10.1016/j.precisioneng.2009.05.007]

[51]   D. Meshram, and Y. Puri, "EDM electrodes manufacturing using rapid tooling concept", *International Journal of Engineering Research and Development,* vol. 3, pp. 58-70, 2012.

[52]   P. Balasubramanian, and T. Senthilvelan, *Optimization of machining parameters in EDM process using cast and sintered copper electrodes*, 2014.
[http://dx.doi.org/10.1016/j.mspro.2014.07.108]

[53]   V.K. Pal, and S.K. Choudhury, "Fabrication of texturing tool to produce array of square holes for EDM by abrasive water jet machining", *Int. J. Adv. Manuf. Technol.,* vol. 85, pp. 2061-2071, 2016.
[http://dx.doi.org/10.1007/s00170-015-7875-7]

[54]   P. Goyal, N.M. Suri, S. Kumar, and R. Kumar, "Investigating the surface properties of EN-31 die-steel after machining with powder metallurgy EDM electrodes", *Materials Today: Proceedings,* vol. 4, pp. 3694-3700, 2017.

[55]   A.K. Sahu, S. Chatterjee, P.K. Nayak, and S.S. Mahapatra, "Study on effect of tool electrodes on surface finish during electrical discharge machining of Nitinol", *IOP Conf. Series Mater. Sci. Eng.,* vol. 338, no. 012033, 2018.

[56]   A. Equbal, M.I. Equbal, and A.K. Sood, "An investigation on the feasibility of fused deposition modelling process in EDM electrode manufacturing", *CIRO J. Manuf. Sci. Technol.,* 2019.
[http://dx.doi.org/10.1016/j.cirpj.2019.07.001]

[57]   P. Saxena, and R.M. Metkar, *Development of electrical discharge machining (EDM) electrode using fused deposition modeling (FDM)*, 2019.
[http://dx.doi.org/10.1007/978-981-13-0305-0_22]

CHAPTER 3

Thermally Enhanced Non-Conventional Machining Process- A Review

Janender Kumar[1], Suneev Anil Bansal[1,*] and Munish Mehta[2]

[1] Department of Mechanical Engineering, Maharaja Agrasen University, Baddi, Solan, India

[2] School of Mechanical Engineering, Lovely Professional University, Phagwara, Punjab, India

Abstract: The present work puts light on the non-conventional machining process in which heat energy is utilized to remove the material from the workpiece. Laser beam machining is a process in which a high-intensity beam falls on the work surface to raise the temperature so that the vaporization of material could take place. Laser beam machining process can be utilized to machine hard and brittle materials. The laser types used and laser drilling processes are reviewed. The complex profiles cutting and micro-drilling can be achieved which is not possible by other conventional and non-conventional machining processes. Many researchers have studied that a hybrid process can improve the process parameters of laser beam machining. This paper reviews its need, hybrid process and latest developments in the field of machining of different materials like ceramics, composites, *etc.* The last section of the paper discusses the area of future research in this field.

Keywords: Complex shapes, Drilling, Hard and brittle materials, Heat energy, Hybrid machining, Laser beam, Laser beam Machining, Machining parameters, Micro-drilling.

INTRODUCTION

In the manufacturing industries, the development of advanced material brought a revolution in the performance of equipment and machinery due to its superior quality. Simultaneously, it has posed challenges to its machining because these materials are hard and brittle. The complex shape and unusual workpiece size were also the requirements. Industry has forced the research engineers to develop newer machining methods [8] [12]. So, to take full advantage of these costly materials, advanced machining methods are under consideration. These are ultrasonic machining, electrochemical machining, electrochemical grinding, laser beam machining, water jet machining, plasma arc machining, *etc.* Yet these have

---

* **Corresponding author Suneev Anil Bansal:** Department of Mechanical Engineering, Maharaja Agrasen University, Baddi, Solan, India; E-mail: suneev@gmail.com

own certain limitations regarding machining of parent material. In this paper, laser beam machining and allied processes are studied. It is also called thermally enhanced machining processes because heat energy is utilized for machining. In our childish days, we performed one experiment in which sun rays were focused through a lens to burn a piece of carbon paper. The energy density was about1W/mm^2. In some manners, a laser beam can be focused on hard material with high density to remove the material from the workpiece through vaporization. Einstein gave the concept of laser theory and stimulated emission. The first laser named Ruby laser was produced by Tower and Shawlow in the year 1957.

LASER BEAM MACHINING

The full form of laser is "Light Amplification by Stimulated Emission of Radiation" Laser light is monochromatic in nature *i.e.* its wavelength occupies a very narrow portion of the spectrum. Hence a simple lens can focus the laser light to a micro-spot of a very smaller diameter and with higher intensity than another type of light. Laser light is coherent as it travels in phase. The three major elements of any laser device are:

a. Lasing medium

b. A pumping energy source required to excite these atoms to higher energy levels.

c. Optical feedback system.

The laser medium may be a solid, liquid or gas. All have different wavelengths. It lies between 0.21-11micron meter.

Ruby = 0.7micron meter, Nd:YAG = 1.0 micron meter, CO=2.7micron meter,CO_2=10micron meter.

Types of Lasers

Different types of laser are utilised in machining having a frequency and mean power accordingly as shown in Table **1**.

Difference Between Ordinary & Laser Light

Laser light has photons of the same frequency, wavelength and phase. That's why it is highly directional and of high-power density. It has better focusing properties as compare to ordinary light.

Table 1. Types of laser [3].

Types of Laser	Mean Power	Efficiency
Nd: YAG	1w-2kw	2-5%
CO_2	10w-20kw	10-15%
Excimer	1w-500w	Less than 3
CO	5w-15kw	Less than 20
I_2	Less than 1kw	Less than 20

Principle of Laser Beam Machining

When a laser beam is focused through a lens on workpiece, the heat energy is absorbed which melts the selected surface, melts it, vaporizes it instantaneously and penetrates it. Thus, it can also be called a thermal cutting process as shown in Fig. (**1**).

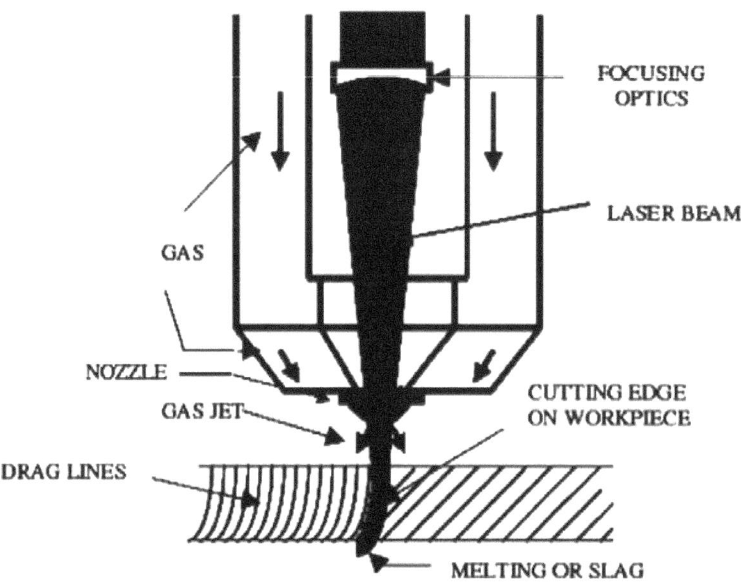

Fig. (1). Laser beam cutting [12].

Laser Drilling Overview

The small size hole, can not be produced by conventional machining processes, can be prepared with laser beam machining. Hence this was the huge advantage for drilling of hard to cut materials. The comparison hole drilling using Laser

Drilling with other processes is shown in Table **2**.

Types of Laser Drilling

There are two types of laser beam drilling process present and these are percussion and trepan drilling. Percussion drilling punches directly through the workpiece material with no relative movement of the laser or workpiece. Trepan drilling involves cutting around the circumference of the hole to be generated as shown in Fig. (**2**).

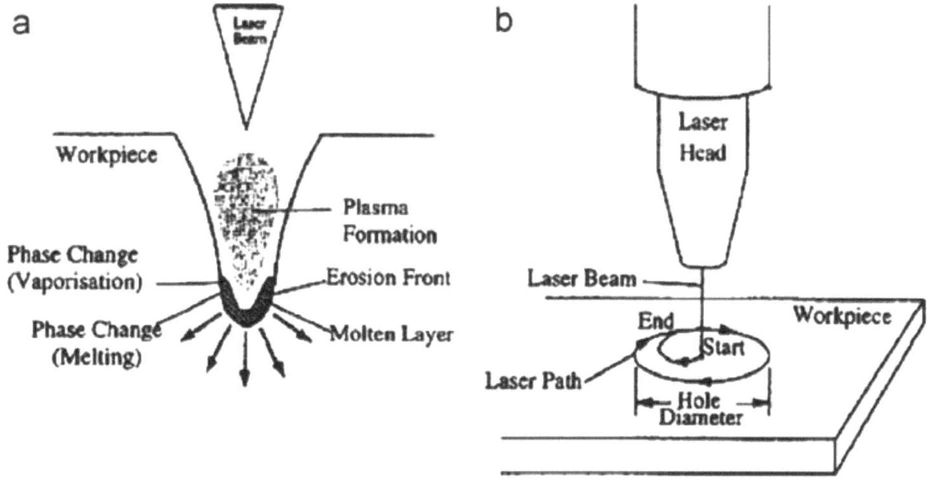

Fig. (2). Schematic of Laser Beam (**a**) Percussion drilling (**b**) Trepan drilling [12].

Table 2. Comparision of non-conventional processes [10].

Non-Conventional Machining Processes	Precision Small Hole (Diameter Less than 0.25mm)
Ultrasonic machining	-
Electrochemical machining	-
Laser beam machining	Good
Plasma arc machining	-
Electrical discharge machining	-
Abrasive jet machining	-
Chemical machining	fair

It is clear from the table that only two processes could produce a hole less than 0.25 in diameters. Laser beam drilling has performed well than chemical machining.

LITERATURE REVIEW

Peyman *et al.* described the use of laser in the medical field. In this Nd: YAG laser energy was utilised to surgery for tumor and it was successful. Now the use of laser is increasing continuously in this direction [1]. Jiang *et al.* discussed about, most wonderful and rapidly increasing, laser beam machining process. The focus was on the use of CO_2 and Nd: YAG laser. Both types of laser were utilised for cutting hard material and had their own limitations, Nd: YAG laser had their low mean beam power but high beam intensity. High beam intensity made it special over CO_2 laser. The maximum depth of machining was a severe problem in pulsed Nd: YAG laser due to low mean beam power. So the study was carried out to estimate the depth of machining and the transverse of the produced hole [2]. Yeo *et al.* investigated the laser drilling operation in the material which was generally used in the aerospace industry. In this paper different, types of the laser having different wavelengths were analysed. The operating power of each laser and their efficiencies were not same . The different parameters of this process were also reviewed for getting good quality holes [3]. Pietrot *et al.* reviewed the laser cutting process and quality in components produced by the laser. The enhancement in quality of a product could be done through proper monitoring, diagnosis and regulation for improvement in cutting. The initial set up and running cost of this process is very high. Proper control must be done to make the process economically viable [4]. Black *et al.* investigated the machining of a thick ceramic tile having thickness of 8.5 and 9.2mm. In this CO_2 laser beam machining was done and different cutting parameters at different cutting speeds studied. With the use of a CO_2 laser, cracks were observed and it was due to temperature gradient (change in temperature with respect to thickness) within the ceramic tiles. To avoid cracks, a cooling arrangement must be incorporated during machining [5]. Black *et al.* investigated the cutting process of ceramic tiles using a CO_2 laser. The various parameters were analysed for preparing the database for this process. Six types of Si/Al_2O_3 based ceramic tiles were considered for analysis having a thickness between 3.7 to 9.2mm. The database was collected for work in two conditions, one from the atmosphere and other from underwater cutting [6]. Fang *et al.* identified that tool wearing was the big problem in diamond turning of glasses. Three methods were discussed in paper for machining glass-like indentation, single point scratching and taper grooving. Out of these, taper grooving was more effective for the study of the transition from brittle to ductile mode. Still, the life of the tool was the major issue [7]. Meijer *et al.* discussed about laser beam machining, its applications with the use of ultrashort pulsed laser. With a short pulsed laser, micromachining could be easily done and this was the advantage of this process. Still, in this field of laser, development is on the way and the research continues in this field [8]. Sun *et al.* summarized the progress and benefits of thermally enhanced machining of hard and brittle materials in which material was removed by the external heat source. It was observed that laser beam machining can improve the machinability in terms of longer tool life, reduction in cutting forces and better surface finish for composite and ceramic materials. It observed that saving in cost of tool and reduction in tool changing time, the life of tool increased in laser and plasma-assisted machining [9] Dhar *et al.* reviewed the laser drilling process and techniques to solve the issues during high quality and accurate holes. The three drilling techniques were discussed. These were single-pulse drilling, trepanning and helical drilling. With these drilling techniques, small to large size holes can be produced. It was concluded that drilling time was more from total time and hence drilling cost was increased. Laser beam drilling has the potential to reduce the drilling time through better control of the process [10]. Pham *et al.* proposed a new technique of machining of alumina material with a laser milling process. Basically, in earlier days, the machining of the ceramic component was very costly because it required costly tooling. In the current paper, the material removal characteristics of this process are reviewed. The results showed that complex geometry can be produced with laser beam milling without the need for expensive tools. So, it concluded that this process was cost-effective if ceramic components were produced in batches [11]. Dubey *et al.* investigated the hybrid processes influence in today's industries. In recent years, laser and non-laser hybrid processes are becoming more popular. Laser and non-laser hybrid conventional machining processes comprises of laser-assisted turning and grinding. Similarly, laser and non-laser unconventional hybrid machining processes comprises of laser-assisted ultrasonic machining, laser-assisted electrochemical and electrical discharge machining. Yue *et al.* found a deeper hole with a much smaller recast layer during ultrasonic-assisted laser drilling as compared with laser-assisted drilling without ultrasonic aid [12] [13]. Low *et al.* reported an investigation of taper hole formation during laser drilling. It was general problem of laser percussion drilling. The taper in hole could be controlled by a laser beam with variable pulses. With identical pulses, the taper hole could be easily seen during drilling. The new approach utilised to produce zero, negative and positive holes [14]. Dubey *et al.* studied the effect of process parameters on the performance of Nd: YAG laser beam machining. The investigation was done with the design of experiments for drilling, micro-drilling and cutting operations. The results showed that Nd: YAG laser beam machining can be successfully applied for machining due to the capability of changing wavelength and frequency. With this process, a micro-hole up to 5-micron meter and the thin sheet of 4-micron meter can be easily machined [15]. Kibria *et al.* experimented with Nd: YAG laser micro turning on the cylindrical workpiece of ceramic material as this has a great advantage in automobile, medical and electronic industries. The depth of cut and surface roughness of material by varying process parameters was analysed. The results showed that process parameters have a great influence on achieving the depth of cut and surface roughness. Optical and SEM micrographs have been analysed for proper knowledge of the laser micro process parametric setting. It found that lamp current 23 amp, pulse frequency 800HZ, scanning speed 6.6 mm/s, good surface finish obtained after single laser beam micro-turning [16]. Kim *et al.* performed the study on the hybrid process, the new process was laser-assisted turning. In this process, laser-assisted machining and CNC lathe were utilised for machining. Three issues were discussed in laser-assisted turning. These were the machining process, heat treatment and device mechanism. In this process, high power diode laser has been utilised instead of CO_2 or Nd: YAG laser. As it has a high absorption rate than others due to its short wavelength. In the field of laser beam turning very small work has been observed, especially for milling. So, it becomes a necessity to research in this field [17]. Phiphon *et al.* utilized the general algorithm to improve and optimize the process parameters of laser beam machining. Experiments and mathematical models are designed using response surface methodology and then general algorithm applied to get optimum results. The best results in terms of surface roughness and kerf taper were obtained. Thus, results proved that the general algorithm is a good optimization technique [18]. Loumena *et al.* discussed the hidden potential of light-weighted CRPF material. Laser processes increase the feasibility of this high strength and good corrosion resistance material through machining. In the experiment, nanosecond laser, based on new generation BOREAS series which have certain technical specification was utilised. With the use of this laser, drawbacks of more heating of CRPF materials reduced up to some level and suggested to use excimer, rod-type fibre laser in the near future [19]. Mishra *et al.* elaborated on the coupled methodology for the development of a prediction model for laser beam percussion drilling. The two methodologies, finite element analysis and the artificial neural network, were utilised. The experiment was conducted for hole taper which created during the laser drilling process. The optimised parameters show great improvement in reducing hole taper by 67.5%, reduction of heat affected zone by 3.25% and increases in material removal rate by65% [20]. Nasim *et al.* discussed the diode laser dependent technologies along with measuring techniques. It could produce a very small size having the reliability of machining. The application of the process is in medical for cancer recognition, explosive detection, monitoring of air and water quality. The current paper represented diode laser use from medical laboratory to industries for doing machining [21]. Venkata *et al.* reviewed all non-conventional machining processes and the main attention was on optimization parameters of these processes. The processes are electrochemical machining, micro-machining, nano machining, hybrid processes and laser beam machining. The purpose of this review was to give an overview of all processes and to give direction for new research in this field. The comparison of these with their allied processes was represented in this paper [22]. Venkatesan *et al.* summarized the laser-assisted machining and it's potential to cut hard and brittle materials which was a hybrid method. In this paper, laser assisted machining of titanium alloy, nickel-based alloys, ceramics, composites were studied and left the scope for optimization of laser assisted machining process. It was concluded that the hybrid process can be used to increase the performance of materials which were difficult to machine with conventional methods [23]. Liu *et al.* introduced the ultrasound-assisted water confined laser micromachining. Actually, during laser beam machining, many defects produced and also generated chips was deposited on the workpiece material. It produced harmful thermal residual effects. In this paper ultrasound assisted water confined laser machining was introduced which eliminated the defects which were produced in laser beam machining. The results showed that less deposition of debris observed in this [24]. Mishra *et al.* discussed about ultrashort laser pulses for creating micro-machining in hard materials and its analysis. Industries can utilize this advanced method of machining. It was called laser beam micro-machining whereas, in laser beam machining, high laser pulses were utilized. Laser beam micro-machining is used for making read-only memory chips, 3d sized structure and biological optical chips. The current paper highlighted the nature of short or ultrashort laser pulses, having interaction with different types of materials [25]. Darnish *et al.* compared the two-machining medium, wet and dry which was employed for machining Inconel 718 superalloy. It was found that machining underwater was more productive than dry medium [26]. Ahman *et al.* compared the micro-EDM and laser machining, a hybrid process. It was observed that with this process machining time was reduced between 50-65% without compromising the quality of micro-holes. The micro-holes were drilled in Ni-Ti alloy which was difficult to machine by the conventional machining process. So, this hybrid process was utilized to machine it. Micro-EDM process is utilized for making holes less than 200micron meter. The material removal rate with the brass electrode was relatively higher in micro-drilled hole of Ni-Ti alloy and 40-65% increase observed without compromising the quality of micro-hole [27]. Alahmari *et al.* presented the laser beam machining underwater for the fabrication of micro-channels in nickel-based Inconel 718 super alloy and different parameters of laser analyzed results showed that underwater laser beam machining can improve the quality of workpiece [28]. Lee *et al.* reviewed the laser and arc heat source for machining individually. The laser arc assisted hybrid process was purposed by the author [29]. Firdaus *et al.* reviewed the tribological aspects of the surface properties of the material. Laser beam machining using Nd: YAG laser was used to improve the surface texture by reducing wear and friction. In this paper, surface texturing was compared between two performing ways. Generally, in most methods, texturing was done directly on workpiece and in certain cases, researches conducted it on a coated material. The coated materials contribute to reducing friction and wear. Research showed that ceramic coated material was utilized to improve the surface texture by removing friction and wear [30]. Lee *et al.* reviewed the laser assisted hybrid processes as these processes had a great impact on various hard to cut materials. With the use of these processes, improvement in machining was observed and reduction in cutting forces up to a large extent was identified too. The characteristics of different lasers were discussed and compared. The CO_2 has a higher wavelength [31]. Wang *et al.* demonstrated the drilling of ceramic material with laser beam machining. The quality of the drilled hole was studied and computational approaches were utilized for input parameters of laser beam machining. Different methods of ceramic drilling were discussed and more work is still required in the field of laser drilling to achieve more accuracy [32]. Mistry *et al.* focused the study on liquid -assisted laser beam micro-machining. This was great approach to machine the surface of the material at the micron level. With this process thermal damage, due to high heat and poor surface finish, can be improved up to some extent. To understand the process parameters, the finite element simulation technique was utilised. Results showed that machining underwater gave good results by reducing surface damage and improving surface finish. Micromachining was performed with or without the use of liquid [33]. Dutt *et al.* focused the research on pulsed Nd: YAG laser beam in the drilling of different high strength, hard to cut materials like superalloys, composites, ceramics, ferrous and non-ferrous materials. The review was emphasized on the use of laser for drilling of different material for increasing the productivity of the process [34]. Sherpa and Pradhan explored the various parameters to enhance the micro-machining of different materials using Nd; YAG laser. The diode-pumped solid-state laser was used for micro-machining. The material removal rate was found to increase linearly with an increase in frequency. Surface roughness was also found to increase with respect to diode current [35]. Rana *et al.* studied the different techniques for checking the machining effect on material and reason for change in material properties. Researchers also explored the use of laser beam machining in the industries, especially aerospace, military and aviation. Various techniques were implemented like Taguchi method, genetic algorithm, response surface method. The inter related parameters were observed. The result indicated that during mild steel cutting, oxygen gas gave the best results for machining thicker material than inert gas [36]. Faisal *et al.* reviewed the laser beam machining and laser beam micro-machining processes. These techniques brought a revolution in the field of machining to hard material. The laser process has been utilised in medical, aerospace and communication also. The laser can produce a hole < 0.3 mm in diameter. That's why it is also called laser beam micro-machining as this size cannot be obtained through any other conventional or non- conventional machining process. Metal micromachining includes both micro-drilling and micro-milling. The recent advancement in the field of laser micromachining has been observed [37]. Holmberry *et al.* evaluated the surface condition of a workpiece after machining. Three non-conventional machining processes were considered and compared for surface integrity. The processes were abrasive water jet machining, electro- discharge machining and laser beam machining. The experiment was performed on alloy 718. The result showed that abrasive water jet machining produced good quality surface conditions with low surface roughness as compared to EDM and LBM. This was the main objective of the investigation [38]. Vasanth *et al.* performed experiment by doing machining on buffalo leather. The CO_2 laser was utilised for cutting purpose and it was observed that laser beam machining could reduce the poor quality/waste formation on the leather surface as compare to manual process [39]. Zhai *et al.* performed experiment for Ni based alloy with thermal barrier coatings by percussion drilling in Inconel 718 material with a thickness of 2000-micron meter with femtosecond laser. With this process, zero taper hole was produced on nickel-based alloy having inner diameter 162-micron meter. Nd:YAG solid state femtosecond laser was utilised for percussion drilling. The study provided the feasibility of femtosecond laser for machining film cooling hole of turbine blades [40]. Hofy *et al.* reviewed the research work carried out in machining of carbon fibre reinforced polymer (CRPF) material by laser beam machining. In this paper, various techniques were critically analysed to improve productivity by reducing machining time and increasing material removal rate. With the conventional machining process of the CRPF material, the tool wear rate was higher as compare to non-conventional machining. So, the study showed laser beam machining as an alternative process for CRPF material with better accuracy [41]. Abdo *et al.* evaluated the performance of the laser process in which micro-channels of v & u-shaped cross-sections were obtained. The study continued at two points; one was the geometrical characteristics and other quality of machined micro-channels. The two techniques were utilised ANOVA and SEM. The following observations were noticed. Control laser parameters can produce high value of laser intensity to produce v & u secondly, the surface finish of the channel can be obtained nearly 2-micron meter [42].

The compilation or a quick review on certain process of laser beam and its associated processes are shown in Table **3**.

The present review provides us the base for doing further research in the field of laser beam machining.

During literature review, it has been observed that Nd: YAG laser was utilised by most of researchers in their experiment for machining. So, in near future, study for other types of laser can be the topic for further research.

Nd: YAG Laser Overview

The abbreviation of it is 'Neodymium-doped Yttrium Aluminium Garnet'. It is a powerful solid-state laser. It is used for light work. High reflective and dielectric material can be easily machined. With pulsed mode it can machine even thicker material. It is compact and reliable type laser. This laser has low mean beam power but high energy density when used with pulsed mode. It is used as a laser medium in laser beam machining process. The schematic diagram of process is in Fig. (**3**).

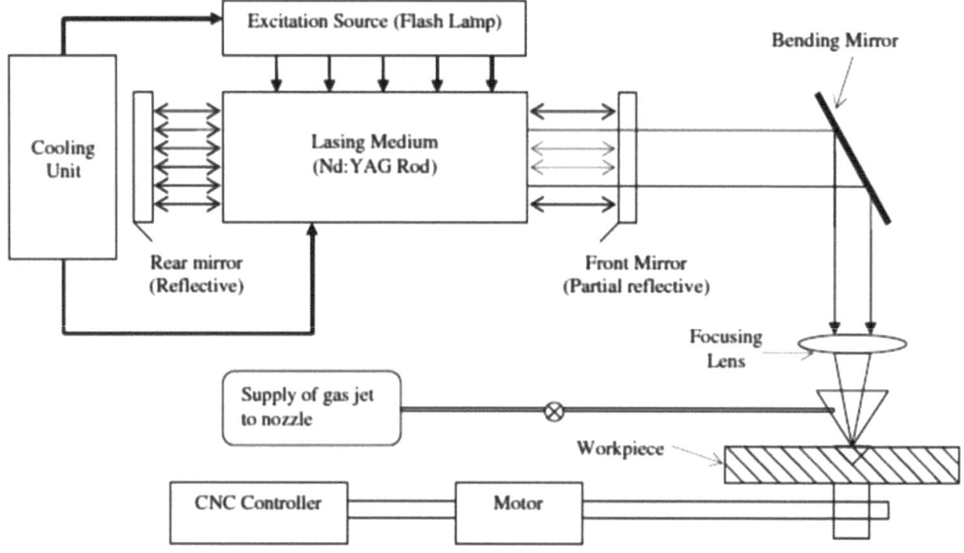

Fig. (3). Schematic of Nd: YAG laser beam cutting process [12].

Table 3. Compilation of laser beam machining and laser-assisted processes [22].

Author	Process	Work Material	Important Input Variables Considered	Optimization Technique	Remarks
Pham *et al* (2007)	Laser milling process	Alumina material and silicon nitride ceramic	Scanning speed. pulse intensity, lamp current	Single factor experiment	It was recommended that for any m/c employing laser milling with the wavelength 1064 nano-meter: a glass support plate should be provided to reposition the workpiece.

(Table 3) cont.....

Author	Process	Work Material	Important Input Variables Considered	Optimization Technique	Remarks
Dube• *et al* (2008)	Laser drilling-micro-drilling	Composites, ceramic and alloys	Laser, materials and process parameters	DOE, Taguchi	YAG laser was suitable for creating micro -holes up to 5micron meter_
Allman- *et al* (2015)	Laser micro-drilling	Ni-Ti alloy	Lamp current, pulse frequency, pulse width, air pressure	Hybrid process	With hybrid process (laser and micro-EDM): machining time reduced and 'RR. increased
Wang *et al* (2016)	Laser drilling process	ceramic	Pulse -width, energy density and repetition frequency	AN YS and COMSOL approaches	It was recommended that liquid assisted processing is good in laser drilling due to cooling &- evaporating process.
Dutt *et al.* (2018)	Laser drilling process	Composite, ceramic and super alloys	Q-switching, peak power, pulse time, pulse frequency, no of pulses and gas pressure	Artificial intelligence-based methods (ANN: Genetic algorithm)	Pulsed Nd: Y..A.G laser beam drilling was mu easing rapidly due to better focussing characteristics

CONCLUSIONS

1. Laser beam machining is utilized for making complex profile in workpiece, drilling holes in different sizes (micro-holes up to 5micron meter in diameter) and cutting of thin foils up to 4-micron meter.

2. The performance of laser depends upon three parameters. These are process, laser and material parameters.

3. The laser hybrid machining processes have produced better results than single non-conventional machining process.

4. Coolant is used in conventional and non-conventional machining during processes. Coolant remains toxic for environment and operator as it continuously used on machines. But laser-assisted machining is an environment friendly process due to the absence of it.

5. Laser-assisted machining is better than conventional machining. Also, it can be used to increase the process efficiency of material those tough to machine by

traditionally method.

6. It is concluded that the use of Nd: YAG laser, machining of Inconel 718 superalloys using laser beam machining under wet conditions with the static mode of water, has given best results in terms of maintaining the sizes, burr free surface and recast layers.

7. It is also concluded that hard to cut material can be easily drilled by using pulsed Nd: YAG laser beam drilling. This process, practically and theoretically, may be carried out successfully for composites. This application of laser is increasing continuously.

FUTURE SCOPE FOR LASER BEAM MACHINING RESEARCH

1. Most research papers are related to drilling and micro-drilling. Still, the three-dimensional laser beam machining like turning and milling are difficult to perform. Research can be carried out for laser assisted milling in near future.

2. The laser beam machining is generally performed on sheet metals but due to advanced or later engineering material need arises to develop it for machining of difficult to hard materials. So, this can be the research area in future.

3. During laser process, in front of tool, workpiece remains soft due to external heat source. Also, the tool and chip interface have high temperature which can leads to shorten tool life. Therefore, a proper method for cooling the cutting tool without affecting the heat up area in front of tool is required [9].

4. Generally, this process is utilized for hard and brittle materials. But some ductile materials are also difficult to machine. Hence thermally enhanced machining may be utilized for ductile material for improving its machinability [9].

5. For optimization of thermally enhanced machining process, a depth study of tribological elements like friction, wearing and rubbing between cutting tool and chips are required.

CONSENT FOR PUBLICATION

Not applicable.

CONFLICT OF INTEREST

The authors confirm that this chapter content has no conflict of interest.

ACKNOWLEDGEMENTS

The authors would like to express their sincere thanks to the editor and anonymous reviewers for their time and valuable suggestions.

REFERENCES

[1]   G.A. Peyman, A. Alghadyan, and J.H. Peace, "A contact Nd:YAG laser to resect large ciliary body and choroidal tumors", *Int. Ophthalmol.,* vol. 11, no. 1, pp. 55-61, 1987.
[http://dx.doi.org/10.1007/BF02027898] [PMID: 3692695]

[2]   C.Y. Jiang, W.S. Lau, T.M. Yue, and L. Chiang, "On maximum depth and profile of cut in pulsed Nd: YAG laser machining", *Annals of CIRP,* vol. 42, no. 1, pp. 223-226, 1993.
[http://dx.doi.org/10.1016/S0007-8506(07)62430-5]

[3]   C.Y. Yeo, S.C. Tam, S. Jana, and M.W.S. Lau, "A technical review of laser drilling of aerospace material", *J. Mater. Process. Technol.,* vol. 42, pp. 15-49, 1994.
[http://dx.doi.org/10.1016/0924-0136(94)90073-6]

[4]   P. Di Pietrot, and Y.L. Yao, "An investigation into characterizing and optimizing laser cutting quality - A review", *Int. J. Mach. Tools Manuf.,* vol. 34, no. 2, pp. 225-243, 1994.
[http://dx.doi.org/10.1016/0890-6955(94)90103-1]

[5]   I. Black, and K.L. Chuck, "Laser cutting of thick ceramic tile", *Opt. Laser Technol.,* vol. 29, no. 4, pp. 193-205, 1997.
[http://dx.doi.org/10.1016/S0030-3992(97)00005-4]

[6]   I. Black, S.A.J. Livingstone, and K.L. Chua, "A laser beam machining database for the cutting of ceramic tile", *J. Mater. Process. Technol.,* vol. 84, pp. 47-55, 1998.
[http://dx.doi.org/10.1016/S0924-0136(98)00078-8]

[7]   F.Z. Fang, X.D. Liu, and L.C. Lee, ""Micro-machining of optical glasses - A review of diamond-cutting glasses," Sadhana - Acad", *Proc. Eng. Sci.,* vol. 28, no. 5, pp. 945-955, 2003.

[8]   J. Meijer, "Laser beam machining, state of art and new opportunities", *J. Mater. Process. Technol.,* vol. 149, pp. 2-17, 2004.
[http://dx.doi.org/10.1016/j.jmatprotec.2004.02.003]

[9]   S. Sun, M. Brandt, and M.S. Dargusch, "Thermally enhanced machining of hard-to-machine materialsA review", *Int. J. Mach. Tools Manuf.,* vol. 50, no. 8, pp. 663-680, 2006.
[http://dx.doi.org/10.1016/j.ijmachtools.2010.04.008]

[10]   S. Dher, and N. Saini, "A review on laser beam drilling and its techniques", International conference on advances in mechanical engineering, BBSBEC, Pb",

[11]   D.T. Pham, S.S. Dimov, and P.V. Petkov, "laser milling of ceramic components", *Int. J. Mach. Tools Manuf.,* vol. 47, pp. 618-626, 2007.
[http://dx.doi.org/10.1016/j.ijmachtools.2006.05.002]

[12]   A.K. Dubey, and V. Yadava, "Laser beam machining-a review", *Int. J. Mach. Tools Manuf.,* vol. 48, pp. 609-628, 2008.
[http://dx.doi.org/10.1016/j.ijmachtools.2007.10.017]

[13]   T.M. Yue, T.W. Chan, H.C. Man, and W.S. Law, "Analysing ultrasonic aided laser drilling using infinite element method", *CIRP,* vol. 45, no. 1, pp. 169-172, 1996.
[http://dx.doi.org/10.1016/S0007-8506(07)63040-6]

[14]   "Li.l.low, D.K.Y. Ghoreshi and J.R. Crookall," Hole taper characterisation and control in laser percussion drilling", *CIRP Ann.,* vol. 51, no. 1, pp. 153-156, 2002.
[http://dx.doi.org/10.1016/S0007-8506(07)61488-7]

[15] A.K. Dubey, and V. Yadava, " experimental study of Nd: YAG laser beam machining-a review"'", *Journal of material processing technology,* vol. 195, pp. 15-26, 2008.

[16] G. Kibria, B. Doloi, and B. Bhattacharyya, "Experimental analysis on Nd:YAG laser micro-turning of alumina ceramic", *Int. J. Adv. Manuf. Technol.,* vol. 50, no. 5–8, pp. 643-650, 2010.
[http://dx.doi.org/10.1007/s00170-010-2527-4]

[17] K.S. Kim, "A review on research and development of laser assisted turning"'", *International journal of precision and manufacturing,* vol. 12, pp. 753- 759, 2011.

[18] R. Phipon, and B.B. Pradhan, "control parameters optimisation of laser beam machining using genetic algorithm", *International journal of computational engineering research,* vol. 2, no. 5, pp. 1500-1516, 2012.

[19] C. Loumena, M. Nguyen, J. Lopez, and R. Kling, "Potentials for lasers in CFRP production",
[http://dx.doi.org/10.2351/1.5062378]

[20] S. Mishra, and V. Yadava, *Modelling and optimisation of laser beam percussion drilling of thin aluminium sheet", Optics and laser technology.* vol. Vol. 48. , 2013, pp. 461-474.

[21] H. Nasim, and Y. Jamil, "Diode lasers: From laboratory to industry", *Opt. Laser Technol.,* vol. 56, pp. 211-222, 2014.
[http://dx.doi.org/10.1016/j.optlastec.2013.08.012]

[22] R. Venkata, "optimisation of modern machining processes using advanced optimisation techniques-a review", *International journal of advanced manufacturing technology,* vol. 73, no. 5-8, pp. 1159-1188, 2014.

[23] K. Venkatesan, and R. Ramanujan, *Procedia Eng.,* vol. 97, pp. 1626-1636, 2014.
[http://dx.doi.org/10.1016/j.proceng.2014.12.313]

[24] "Z. liu,Y. Gao, B. Wu, N. Shen, and H. Ding, "Ultrasound-assisted water-confined laser micromachining: A novel machining process", *Manuf. Lett.,* vol. 2, no. 4, pp. 87-90, 2014.
[http://dx.doi.org/10.1016/j.mfglet.2014.06.001]

[25] "finite element analysis and simulation of liquid assisted laser beam machining", *Int. J. Adv. Manuf. Technol.,* vol. 94, no. 5-8, pp. 2325-2331, 2017.

[26] S. Darwish, N. Ahmed, A.M. Alahmari, and N.A. Mufti, "A comparison of laser beam machining of micro-channels under dry and wet medium", *Int. J. Adv. Manuf. Technol.,* vol. 83, pp. 1539-1555, 2015.
[http://dx.doi.org/10.1007/s00170-015-7658-1]

[27] A. Ahman, M.S. Rasheed, M.K. Mohammed, and T. Saleh, "A hybrid machining process combining micro-machining and laser beam machining of Ni-Ti based shape memory alloys", *Mater. Manuf. Process.,* vol. 31, no. 4, pp. 447-455, 2015.

[28] A.M. Alahmari, N. Ahmed, and S. Darwish, "laser beam micromachining under water immersion", In: *International journal of advanced manufacturing technology* vol. 83. , 2016, pp. 1671-1681.

[29] C.M. Lee, W.S. Woo, D.H. Kim, and W.J. Oh, "laser assisted hybrid processes-a review", *International journal of precision engineering and manufacturing,* vol. 17, pp. 257-267, 2016.

[30] A. Firdaus, S. Baharin, M.J. Ghazil, and J.A. Wahab, "laser surface texturing and its contribution to friction and wear reduction-a brief review", *Ind. Lubr. Tribol.,* vol. 8, no. 1, pp. 57-66, 2016.

[31] C.M. Lee, W.S. Woo, J.T. Baek, and E.J. Kim, "laser and arc manufacturing process-a review", *Int. J. Precis. Eng. Manuf.,* vol. 17, no. 7, pp. 973-985, 2016.
[http://dx.doi.org/10.1007/s12541-016-0119-4]

[32] H. Wang, H. Lin, C. Wang, L. Zheng, and X. Hu, "Laser drilling of structural ceramics—A review", *Journal of Europian. Ceram. Soc.,* vol. 37, no. 4, pp. 1157-1173, 2016.
[http://dx.doi.org/10.1016/j.jeurceramsoc.2016.10.031]

[33]　V. Mistry, and S. James, "finite element analysis and simulation of liquid assisted laser beam machining process", *Int. J. Adv. Manuf. Technol.,* vol. 94, no. 5-8, pp. 2325-2331, 2017.
[http://dx.doi.org/10.1007/s00170-017-1009-3]

[34]　G.D. Gautam, and A.K. Pandey, "Pulsed Nd: YAG laser beam drilling-a review", In: *Optics and laser technology* vol. 100. , 2018, pp. 21-1833.

[35]　T.D. Sherpa, and B.B. Pradhan, "Micro-grooving of silicon wafer by Nd:YAG laser beam machining", *IOP Conf. Series Mater. Sci. Eng.,* vol. 377, pp. 1-7, 2018.
[http://dx.doi.org/10.1088/1757-899X/377/1/012219]

[36]　R.S. Rana, R. Chouksey, K.K. Dhakad, and D. Paliwal, "Optimization of process parameter of Laser beam machining of high strength steels: A review", *Mater. Today Proc,* vol. 5, no. 9, pp. 19191-19199, 2018.
[http://dx.doi.org/10.1016/j.matpr.2018.06.274]

[37]　N. Faisal, D. Zindani, K. Kumar, and S. Bhowmik, "laser micromachining of engineering materials and tribology-a review", *Material forming, machining and tribology,* pp. 121-136, 2018.

[38]　J. Holmberg, and A. Berglund, "Wretland and T. Bano", Evaluation of surface integrity after high energy machining with EDM, laser beam machining and abrasive water jet machining of alloy 718", *Int. J. Adv. Manuf. Technol.,* vol. 100, pp. 1575-1591, 2018.
[http://dx.doi.org/10.1007/s00170-018-2697-z]

[39]　S. Vasanth, and T. Muthuramalingam, "a study on machinability of leather using co2-based laser beam machining process", *Advances in manufacturing processes,* pp. 239-244, 2018.

[40]　Z. Zhai, and X. Wang, "percussion drilling on nickel-based alloy with thermal barrier coatings using femtosecond laser", *International journal for light and electron optic,* vol. 194, p. 163066, 2019.

[41]　M.H. Ei-Hofy, and H. Ei-Hofy, "laser beam machining of carbon fibre reinforced composites-a review", *Int. J. Adv. Manuf. Technol.,* vol. 101, no. 9-12, pp. 2965-2975, 2019.
[http://dx.doi.org/10.1007/s00170-018-2978-6]

[42]　B.H.A. Abdo, N. Ahmed, A.M. El-Tamimi, S. Anwar, H. Alkhalefah, and E.A. Nasr, "laser beam micro-machining of zirconia ceramic-a review",an investigation of micromachining geometry and surface roughness", *J. Mech. Sci. Technol.,* vol. 133, no. 4, pp. 1817-1831, 2019.
[http://dx.doi.org/10.1007/s12206-019-0334-x]

CHAPTER 4

AWJM Process- A Review

Mamta [*,1], **Sachin Mohal**[2], **Saurabh Chaitanya**[2] and **Ankitmani Tripathi**[3]

[1] *Department of Mechanical Engineering, Maharaja Agrasen University, Baddi, Solan, India*

[2] *Department of Mechanical Engineering, Chandigarh Engineering College, Landran, Mohali, India*

[3] *Mechanical Engineering Department, SRM University Delhi NCR Haryana, India*

Abstract: Abrasive water jet machining (AWJM), non-conventional machining (NCM), has been improved radically, in terms of quality and reliability. AWJM is a widely used NCM process broadly established in technical application areas of machining tools, parts, microstructures or intricate products like Titanium Alloy, Alloy steel *etc.* It has become entrenched in every single significant region of theoretical researches and across the broad spectrum of applications. Applications include medical apparatuses, precision parts, efficient material, smart automotive manufacturing, aerospace equipment, renewable energy science *etc.* This paper audits the verifiable, most recent research improvements, incorporated uses of AWJM system and execution. The examination encourages the high assembling accuracy and balanced out execution quality with more noteworthy productivity. The benefits of AWJM incorporates the capacity to cut convoluted materials, difficult to cut surfaces, high flexibility, little cutting powers and ecologically safe. Additionally,the AWJM system's limitations are highlighted too. AWJM is one of the exceptionally advanced system that have significant potential for micromachining.

Keywords: Abrasive water jet, Machining, Materials.

INTRODUCTION

A fast turn of events, in designing and innovation, needs exceptionally exact and profoundly précised items. Surface properties play a significant role in creating products for automotive, aviation and biomedical applications. A little scratch or burr may cause huge misfortune, for example, midway engine failure, aviation catastrophe and so on. Industries are endeavoring hard and spending gigantic sum on completing of these, without harming surface property. It is a difficult task machine hard and tough material without changing its surface quality and surface property. In NCM, the heat is generated during machining which may prompt sur-

---

[*] **Corresponding author Mamta:** Department of Mechanical Engineering, Maharaja Agrasen University, Baddi, Solan, India; E-mail: dahiya.mamta56@gmail.com

face deformities [1]. Along these lines, by choosing the suitable machining process, we can give better surface quality. In this specific situation, a water jet can be used to machine the hard and tough materials without the generation of any heat. This mix *i.e.* water and abrasive, called as water jet machining (WJM), is a NCM procedure where high speed stream of water is utilized to erode materials from the outside of the work piece. WJM can likewise be utilized to cut soft materials like plastic, elastic or wood. Be that as it may, to cut more hard materials like metals or rock, an abrasive material is blended in the water. At the point when an abrasive material is utilized, in water for the machining procedure, it is called as AWJM. It works on water erosion. At the point when the stream of water with high speed strikes the surface, the expulsion of material happens. Dr. Norman C. Franz started the advanced WJM innovation in 1968, which was granted the principal patent for a creative high-pressure WJM framework and in 1971 the main business water jet cutting was created to cut the overlaid paper tubes. In 1980, Dr. Hashish understood the way toward adding abrasives to plain WJM and from that point AWJM was invented, to cut such mechanical materials, for example, steel, glass, and cement [2, 3].

In the WJM process, a high speed water jet is made to strike the work piece. Pressure force applied on the water, at the end of the nozzle, converted into high velocity. Later on, to cut the material of high quality and to cut hard materials, abrasives are added to high velocity water jet. Several advantages are associated with WJM and cost structure is less too. This innovation is the rapidly developing and flexible. Poisonous exhaust, recast layers, slag and residual stresses can be easily eliminated. Cutting powers are less in AWJM and there is no heat influencing zone close in the vicinity of cutting zone as present in NCM. In this way, it has been demonstrated by that the materials which cannot be machined easily by NCM can be effectively machined by WJM, for example, cutting slots in glass, stone, and metals, fast boring of opening in titanium *etc.* can be effortlessly machined by WJM [4].

(Fig. **1**) presents a schematic image of AWJM components. The significant segments of the AWJM process are high pressure liquid pump (HPLP), high pressure pump, computerized control system, head for machining and catcher. The HPLP is utilized for providing to ultrahigh pressure, consistently.The system comprises of two tanks that cater to the requirement of cutting material and water.The Whole system is equipped with a filtration mechanism that can filter out larger particles. Large particle in the system can choke the working of the whole system [6].

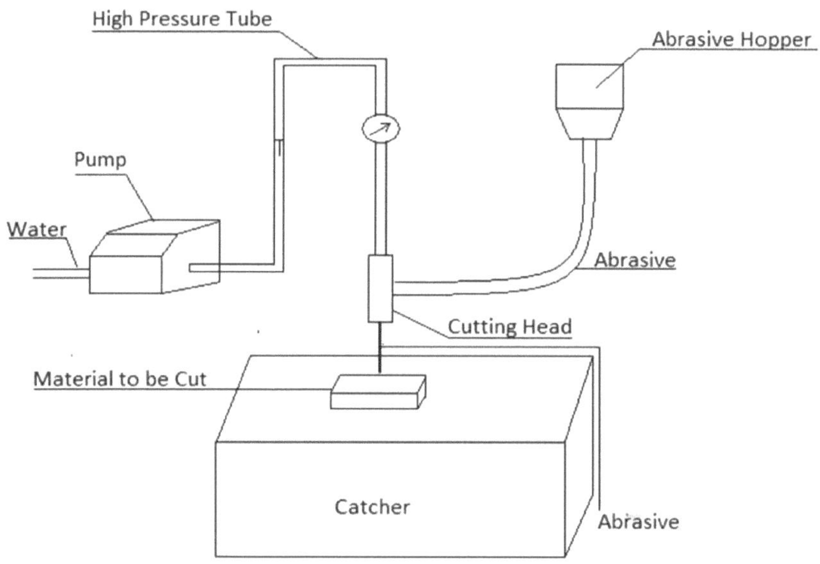

Fig. (1). Schematic illustration diagram of the AWJM process with various components and its attachment [5].

A wide range of materials, Inconel, Titanium, glass, ceramics, composites, heat-sensitive alloys, *etc.* are shaped for different applications with the AWJM process. The demand for heat resistant and higher strength materials is increasing particularly in aerospace industries. However, these materials are often difficult to machine due to their physical and mechanical properties such as high strength and low thermal conductivity, which requires very high cutting energy and makes the cutting forces and cutting temperature very high, and it even leads to a short tool life. This opens an avenue of Thermally Enhanced Machining (TEM) for hard-t--machine materials. TEM may use an external heat source to heat and soften the work piece. As a result, the yield strength, hardness and strain hardening of the work piece reduces and deformation behavior of the hard-to-machine materials changes to allow the plastic deformation. This enables the materials to be machined easily which are difficult for the machine along with low energy requirement that leads to an increase in material removal rate (MRR) and productivity [7].

Various features and developed machines of the AWJM process have been used to cut wide variety of materials such as hard materials like glasses, hard metal like titanium and Inconel, composites and alloys. The interest of heat resistance and higher quality materials is expanding especially in the field of aviation applications. These materials are hard to machine because of their physical and mechanical properties, for example, high strength and low conductivity, which requires extremely high cutting forces and makes the cutting powers and cutting temperature exceptionally high, . It even prompts a short service life of cutting materials. This opens a road of thermally improved machining (TIM) for difficult to-machine materials. TIM may utilize an external heat source to heat and soften the work piece. Thus, the yield strength, hardness, and strain hardening effect of the work piece decrease. By doing this, difficult to-machine materials changes and permits the plastic deformation. This empowers machine tool for the materials to be machined effectively which are hard to machine alongside low energy requirement that prompts increment in MRR and profitability [7].

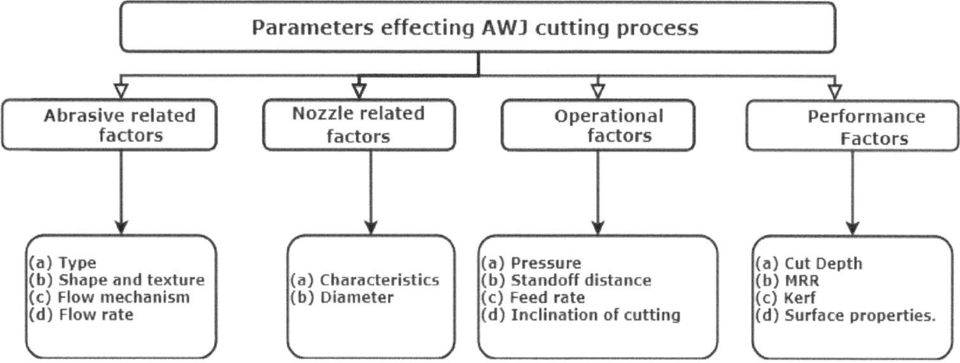

Fig. (2). Diagram showing various classes of parameters affecting the AWJ cutting process [8].

The Performance of thermally improved AWJM (TIAWJM) depends upon various factors (Fig. **2**). These factors can be categorized into various classes. These categories and factors are listed below.

(1) Abrasive related factors

- Type
- Shape and texture
- Flow mechanism
- Flow rate

(2) Nozzle related factors,

- Characteristics
- Diameter

(3) Operational factors

- Pressure
- Standoff distance
- Feed rate
- Inclination of cutting

(4) Performance Factors

- Cut Depth
- MRR
- Kerf
- Surface properties.

A Key role in the machining process is performed by abrasive particles in AWJM. Abrasive, with high velocity, strikes on to the workpiece resulting in erosion of material from the workpiece. This high velocity to the abrasive particle is imparted bythe water jet. Due to the nature of AWJM process, it is capable to cut almost any material [9]. During machining using AWJM the MRR, surface roughness and productivity are related to each other.

These parameters decide the surface quality and determine the functional performance of the resulting product. Geometric accuracy is the major output parameters in the AWJM process too.

By considering about the mechanical properties and conduct, the material of the work piece can be characterized into two groups *i.e.* ductile materials and brittle materials. Both group of materials have a differ mechanism of machining.

Mechanism of machining of ductile materials can be classified into two groups *i.e.* micro-cutting and plastic deformation as shown in Fig. (3) [6].

Fig. (3). Impact of the abrasive particles on a ductile material [6].

Brittle materials, on the other hand, depict a different cutting behavior. The plastic deformation zone and brittle facture are experienced by brittle material while undergoing a cutting process (Fig. **4**).

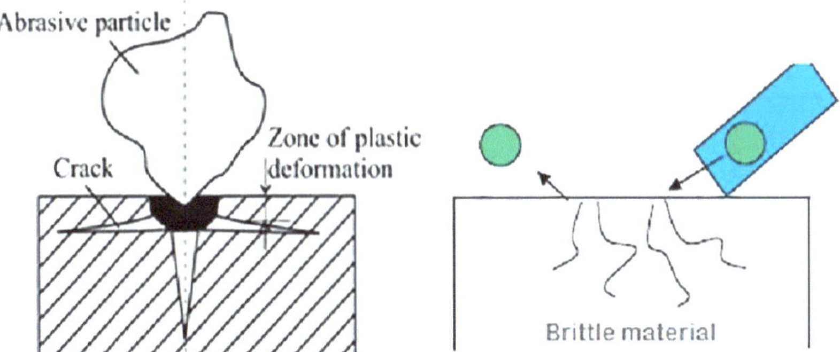

Fig. (4). Impact of the abrasive particles on a brittle material [6].

Several researchers work in this field, and the contribution of Hashish and co-workers is notable. AWJs saw is exceptionally powerful in metal cutting, cutting airplane composites, metals, circuit loads up, and so forth.

AWJM - LITERATURE

Researchers studied AWJM while machining Al7475. Various parameters were considered in the study like pressure, traverse speed, distance from the workpiece and jet inclination. 3mm Sheets of Al were used in the study. Fatigue life and MRR during machining were studied [11]. Yue *et al.* [12] experimented to find the effect of input parameters on the AWJM process. Pressure, feed, abrasive rate and nozzle angle were varied to find the effect on MRR. While cutting alumina, it was concluded that by increasing pressure and abrasive flow, simultaneously, MRR was increased. Numerical simulation studies were performed too with a special type of AWJM process [13]. An ultrasonic assisted AWJM process was investigated to find out the abrasive erosion process. Partially overlapping conditions and entire-overlapping conditions were studied.The Study concluded that a 30° impact angle having a sharp edged particle could lead to higher MRR. Higher MRR was witness at vibration assisted machining rather than without vibration machining [13].

Special materials like Ti6Al4V-CFRP were cut with AWJM [14]. For the first time, taper angle, kerf and roughness were considered during the study. Conventional machining finds various difficulties to cut these materials as both materials (Ti6Al4V and CFRP) have very different characteristic.The Study was successfully performed and opened a new door to machine dissimilar materials in combination widely used in the aerospace industry [14]. Uthayakumar *et al.* studied AWJM to machine supper alloy.The Experiment revealed that AWJM could be successfully used to machine nickel alloy. Precise surfaces could be produced in a cost effective manner using AWJM. The Surface properties of the materials were not degraded during the machining processes. It was observed that pressure and speed during machining can impact kerf wall inclination. It was found that MRR was strongly influenced by parameters like flow velocity and pressure. A high surface finish of super alloy was obtained at high pressure with adequate transverse speed [15]. Temperature was also considered as influencing parameter in AWJM [7]. Thermal treatment at desired locating was used to preheat the workpiece along the path of machining. Preheating was postulated to soften the material that was helpful to substantially reduce the cutting force and increasing the MRR. It was observed that it was very difficult to maintain temperature for a longer time and size. Due to this, it was only valuable for micro cutting and micro channel cutting [7]. Hard to cut stainless steel was also experimented to be cut using AWJ cutting process [16]. EN1 SS was cut using AJM cuttiing. 20mm bar was cut at different presure of water ranging from 3400-3800 bar.

In 2016, Shukla *et al.* optimized the AWJM process parameters for AA6351 Al alloy material. Various optimization technique were used [17].The Study was focused to identify the domain of process parameters with desired output results. Prime facie, the Taguchi method was used to optimize processes parameters that revealed a set of various processes parameters required to obtain desired results [17]. Tamannaee *et al.* contemplated the erosion rate in AWJM. A powder filled thermoplastic olefin was used in machining. From that point, the erosion rate estimations were utilized to create machining configurations to expand the slope of bounding sidewalls [18]. Al-MMCs were also studied to machine with the AWJM process. The Focus of the study was to improve kerf properties [19]. Ti-6Al-4V was also attempted to be machined using AWJM.. Optimizations were also used like Taguchi and ANOVA analysis. A set of experiments were performed to optimize the variable input and to reach the desired output. Desired results output parameters like MRR and kerf properties were optimized [20]. Carbon epoxy, a widely used light weight material, still find limited applications of AWJM. . Therefore, a study was performed to machine carbon epoxy composite material using AWJM [21]. Titanium and its alloys are classified as difficult to cut materials. M. Douiri *et al.* machined Ti–6Al–4V, a titanium alloy, using the AWJM process [22]. The Study revealed several distinct zones while machining. Zone 1, Initial stage damage occurs at lower attack angles. Then cutting progressed to zone 2. The Zone had smooth cutting. The Attack angle was increased. In Zone 3, an erratic cutting was observed. Jet was rather deflected in zone 3 [22]. Inconel finds various applications in engines due to its mechanical properties. The Same mechanical properties make Inconel, a material, difficult to cut. AWJM was successfully used to cut Inconel [23]. To futher improve the study, the Taguchi-Grey relational approach was used to optimize the process parameters [23].

Surface properties of AWJM compared to milling machining were studied too Table **1**. It was concluded that AWJM machine appropriately while generating required finish in a better manner than milling [23]. Suresh *et al.* used the AWJM machining process to drill fiber reinforced composite. Contrary to other reports, abrasive particle size did not influence MRR. It was also observed that pressure influenced the taper of the workpiece [24]. Recently in 2018, Xiaochu *et al.* performed various experiment to develop WJM. Various input and output parameters were analysed [25].

Various theoretical studies were performed too, to analyse WJM Table **2**. Melentiev *et al.* used a theoretical approach to analyze micro AJM. Special attention was given to use small nozzles [26]. Jagadish *et al.* used a soft computing approach. Surface quality and service life was improved using the soft computing approach [27].

Table 1. Review on AWJM.

S.No	Author	Process	Contribution
1.	Haghbin *et al.* [28]	AWJ milling	• Surface evolution model. • Micro channels Machining • Machining relatively deep slots
2.	Jafar *et al.* [29]	AWSJ cutting	• Machining borosilicate glass • MMR • Surface improvement.
3.	Pang *et al.* [2]	AWJ cutting	• Machining amorphous glass • Investigated the channel formation • Predictive model for MRR • Dimensional Prediction
4.	Tamannaee *et al.* [18]	Water jet cutting	• Machining thermoplastic olefin • Machining model pockets
5.	Xu *et al.* [30]	AWJ grinding, lapping, and polishing	• Machining of ceramics • Machining of glasses • Machining of hard–brittle materials
6.	Feng *et al.* [31]	AWJ drilling	• Drilling Small dimension holes • Expediently drilling
7.	Li *et al.* [32]	Radial-mode AWJ turning	• Radial mode AWJ turning • Optimization of various parameters

(Table 1) cont.....

S.No	Author	Process	Contribution
8.	Haghbin *et al.* [33, 34]	WJM	• Experimental WJM • Effect of fluid during machining • Shallow channel machining.
9.	Rupam *et al.* [35]	AWJ drilling	• AWJ drilling • Effect of pressure and time • Controlling the flow rate of abrasive.

Table 2. Overview on the dynamic simulation for WJM process.

S.No	Author	Process	Contribution
1.	Yang *et al.* [36]	COMSOL simulated	• Cathode working zone • Cathode structure optimization
2.	Anwar *et al.* [37, 38]	3D surface generations	• Predictive model • Tuning Strain rate • Heating and friction behavior
3.	Hou *et al.* [39]	Water jet (WJ)	• Velocity vibration analysis • Ultrahigh pressure environment • Flow velocity vibration
4.	Thomas *et al.* [40]	AWJ cut edges	• Formation of aggregate-induced notch cavity • Edge profiles • Plastic deformation behavior

APPLICATIONS OF AWJ MACHINING

The Following are the application of AWJ machining:

1. AWJM process is applicable in the machining of industrial components.

2. AWJM substantially reduces the cycle time and become a potential candidate to replace conventional milling.

3. AWJ cutting has the capability to cut complex shapes and drill difficult holes. Various applications in the cutting of materials have been developed [4].

4. AWJM find application in the field of shale oil to cut hard rocks with the use of abrasive slurry.

5. Applicable in machining heat resistant and higher strength materials use in aerospace.

6. Glass, ceramic, stone, ferrous and nonferrous alloys are successfully cut using AWJM and WJM [41].

7. Sustainable AWJM find applications in fields of sustainable engineering where abrasive and other materials used in cutting are suitably recycled [42].

8. Ti/Ni alloys for aerospace parts are suitably cut using AWJM that are otherwise difficult to cut [43].

9. Sustainable cleaning and degreasing of discs and PCB can be achieved using Ice jet machining [44].

FUTURE SCOPE

Machinability study on stainless steel still is a key area where AWJ can find various applications. There is scope to conduct full factorial experiments. Limited reports are available on surface texture and behavior while machining composite materials. Abrasive particle size and shape distribution are still not fully explored. Future research can see directions in these areas.

CONCLUSIONS

Based on the literature review it has been concluded that WJM methods possess tremendous potential to machine or cut the work piece material. Abrasive jet (AWJM) and water jet (WJM) are extensively used to machine or cut precision part and difficult surfaces. MRR in the AWJM process is most influenced by the abrasive particle size. MRR substantially depends upon carrier mode and fluid properties. The performance characteristics of the AWJM process such as MRR and surface roughness have been improved together by using various optimization techniques. Striking advantages have ben achieved due to the non-existence of heat affected zone. Due to this material do not experience any distortion.

CONSENT FOR PUBLICATION

Not applicable.

CONFLICT OF INTEREST

The authors confirm that this chapter content has no conflict of interest.

ACKNOWLEDGEMENTS

The authors would like to express their sincere thanks to the editor and anonymous reviewers for their time and valuable suggestions.

REFERENCES

[1]   A.S. Walia, V. Srivastava, V. Jain, and S.A. Bansal, "Effect of TiC reinforcement in the copper tool on

roundness during EDM process", *Lect. Notes Mech. Eng.,* pp. 125-135, 2020.

[2]   K.L. Pang, T. Nguyen, J.M. Fan, and J. Wang, "Modelling of the micro-channelling process on glasses using an abrasive slurry jet", *Int. J. Mach. Tools Manuf.,* vol. 53, no. 1, pp. 118-126, 2012.
[http://dx.doi.org/10.1016/j.ijmachtools.2011.10.005]

[3]   G.A. Escobar-Palafox, R.S. Gault, and K. Ridgway, "Characterisation of abrasive water-jet process for pocket milling in Inconel 718", *Procedia CIRP,* vol. 1, no. 1, pp. 404-408, 2012.
[http://dx.doi.org/10.1016/j.procir.2012.04.072]

[4]   S.B. Supriya, and S. Srinivas, "Machinability studies on stainless steel by abrasive water jet-review", *Mater. Today Proc.,* vol. 5, no. 1, pp. 2871-2876, 2018.
[http://dx.doi.org/10.1016/j.matpr.2018.01.079]

[5]   M. C. P. Selvan, and N. M. S. Raju, "A machinability study of stainless steel using abrasive waterjet cutting technology",

[6]   S. Saravanan, V. Vijayan, S.T.J. Suthahar, A.V. Balan, S. Sankar, and M. Ravichandran, "A review on recent progresses in machining methods based on abrasive water jet machining Mater", *Today Proc.,* vol. no. xxxx, pp. 5-11, 2019.

[7]   D. Patel, and P. Tandon, "Experimental investigations of thermally enhanced abrasive water jet machining of hard-to-machine metals", *CIRO J. Manuf. Sci. Technol.,* vol. 10, pp. 92-101, 2015.
[http://dx.doi.org/10.1016/j.cirpj.2015.04.002]

[8]   Y. Gaidhani, "Abrasive water jet review and parameter selection by AHP method", *IOSR J. Mech. Civ. Eng.,* vol. 8, no. 5, pp. 1-6, 2013.

[9]   X. Miao, M. Wu, L. Song, F. Ye, and Z. Qiang, "Research on the method of stacked cutting of abrasive water jet", *Int. J. Adv. Manuf. Technol.,* vol. 103, no. 1–4, pp. 597-604, 2019.
[http://dx.doi.org/10.1007/s00170-019-03567-8]

[10]   A.T. Parameters, "Turning with abrasive-waterjets - A First Investigation", November 1987

[11]   F. Boud, L.F. Loo, and P.K. Kinnell, "The impact of plain waterjet machining on the surface integrity of aluminium 7475", *Procedia CIRP,* vol. 13, pp. 382-386, 2014.
[http://dx.doi.org/10.1016/j.procir.2014.04.065]

[12]   Z. Yue, C. Huang, H. Zhu, J. Wang, P. Yao, and Z. Liu, "Optimization of machining parameters in the abrasive waterjet turning of alumina ceramic based on the response surface methodology", *Int. J. Adv. Manuf. Technol.,* vol. 71, no. 9–12, pp. 2107-2114, 2014.
[http://dx.doi.org/10.1007/s00170-014-5624-y]

[13]   Z. Lv, C. Huang, H. Zhu, J. Wang, P. Yao, and Z. Liu, "FEM analysis on the abrasive erosion process in ultrasonic-assisted abrasive waterjet machining", *Int. J. Adv. Manuf. Technol.,* vol. 78, no. 9–12, pp. 1641-1649, 2015.
[http://dx.doi.org/10.1007/s00170-014-6768-5]

[14]   A. Alberdi, T. Artaza, A. Suárez, A. Rivero, and F. Girot, "An experimental study on abrasive waterjet cutting of CFRP/Ti6Al4V stacks for drilling operations", *Int. J. Adv. Manuf. Technol.,* vol. 86, no. 1–4, pp. 691-704, 2016.
[http://dx.doi.org/10.1007/s00170-015-8192-x]

[15]   M. Uthayakumar, M.A. Khan, S.T. Kumaran, A. Slota, and J. Zajac, "Machinability of nickel-based superalloy by abrasive water jet machining", *Mater. Manuf. Process.,* vol. 31, no. 13, pp. 1733-1739, 2016.
[http://dx.doi.org/10.1080/10426914.2015.1103859]

[16]   Ion aurel Perianu, "Researches regarding the effect of pressure on surface quality during abrasive waterjet cutting of austenitic steels", vol. 4306, 2013

[17]   R. Shukla, and D. Singh, *Experimentation investigation of abrasive water jet machining parameters using Taguchi and Evolutionary optimization techniques.* vol. Vol. 32. Elsevier, 2017.

[18]   N. Tamannaee, J.K. Spelt, and M. Papini, "Abrasive slurry jet micro-machining of edges, planar areas and transitional slopes in a talc-filled co-polymer", *Precis. Eng.,* vol. 43, pp. 52-62, 2016.
[http://dx.doi.org/10.1016/j.precisioneng.2015.06.009]

[19]   K.S.K. Sasikumar, K.P. Arulshri, K. Ponappa, and M. Uthayakumar, "A study on kerf characteristics of hybrid aluminium 7075 metal matrix composites machined using abrasive water jet machining technology", *Proc. Inst. Mech. Eng., B J. Eng. Manuf.,* vol. 232, no. 4, pp. 690-704, 2018.
[http://dx.doi.org/10.1177/0954405416654085]

[20]   K. Nagendra Prasad, D. John Basha, and K.C. Varaprasad, "Experimental Investigation and Analysis of Process Parameters in Abrasive Jet Machining of Ti-6Al-4V alloy using Taguchi Method", *Mater. Today Proc.,* vol. 4, no. 10, pp. 10894-10903, 2017.
[http://dx.doi.org/10.1016/j.matpr.2017.08.044]

[21]   A. Dhanawade, and S. Kumar, "Study on carbon epoxy composite surfaces machined by abrasive water jet machining", *J. Compos. Mater.,* vol. 12, 2018.

[22]   M. Douiri, M. Boujelbene, E. Bayraktar, and S. Ben Salem, "Process reliability of abrasive water jet to cut shapes of the titanium alloy Ti-6Al-4V", *Conf. Proc. Soc. Exp. Mech. Ser.,* vol. vol. 5, 2019pp. 229-236
[http://dx.doi.org/10.1007/978-3-319-95510-0_28]

[23]   M. Kumar, A. Sharma, A.S. Shahi, Á.S.Á. Pwta, and Á. Butt, *Advances in Manufacturing Processes.* Springer Singapore, 2019.

[24]   R. Suresh, K. Sohit Reddy, and K. Shapur, "Abrasive jet machining for micro-hole drilling on glass and GFRP composites", *Mater. Today Proc.,* vol. 5, no. 2, pp. 5757-5761, 2018.
[http://dx.doi.org/10.1016/j.matpr.2017.12.171]

[25]   X. Liu, Z. Liang, G. Wen, and X. Yuan, "Waterjet machining and research developments: A review", *Int. J. Adv. Manuf. Technol.,* 2019.
[http://dx.doi.org/10.1007/s00170-018-3094-3]

[26]   R. Melentiev, and F. Fang, "Theoretical study on particle velocity in micro-abrasive jet machining", *Powder Technol.,* vol. 344, pp. 121-132, 2019.
[http://dx.doi.org/10.1016/j.powtec.2018.12.003]

[27]   S. Jagadish, "Prediction of surface roughness quality of green abrasive water jet machining: a soft computing approach", *J. Intell. Manuf.,* vol. Cm, pp. 1-5, 2015.

[28]   N. Haghbin, J.K. Spelt, and M. Papini, "Abrasive waterjet micro-machining of channels in metals: Model to predict high aspect-ratio channel profiles for submerged and unsubmerged machining", *J. Mater. Process. Technol.,* vol. 222, pp. 399-409, 2015.
[http://dx.doi.org/10.1016/j.jmatprotec.2015.03.026]

[29]   R. Haj Mohammad Jafar, H. Nouraei, M. Emamifar, M. Papini, and J.K. Spelt, "Erosion modeling in abrasive slurry jet micro-machining of brittle materials", *J. Manuf. Process.,* vol. 17, pp. 127-140, 2015.
[http://dx.doi.org/10.1016/j.jmapro.2014.08.006]

[30]   S. Xu, and J. Wang, "A study of abrasive waterjet cutting of alumina ceramics with controlled nozzle oscillation", *Int. J. Adv. Manuf. Technol.,* vol. 27, no. 7–8, pp. 693-702, 2006.
[http://dx.doi.org/10.1007/s00170-004-2256-7]

[31]   Y.X. Feng, C.Z. Huang, J. Wang, X.Y. Lu, and H.T. Zhu, "Surface characteristics of ceramics milled with abrasive waterjet technology", *Key Eng. Mater.,* vol. 329, pp. 335-340, 2007.
[http://dx.doi.org/10.4028/www.scientific.net/KEM.329.335]

[32]   W. Li, H. Zhu, J. Wang, Y.M. Ali, and C. Huang, "An investigation into the radial-mode abrasive waterjet turning process on high tensile steels", *Int. J. Mech. Sci.,* vol. 77, pp. 365-376, 2013.
[http://dx.doi.org/10.1016/j.ijmecsci.2013.05.005]

[33] N. Haghbin, J.K. Spelt, and M. Papini, "Abrasive waterjet micro-machining of channels in metals: Comparison between machining in air and submerged in water", *Int. J. Mach. Tools Manuf.,* vol. 88, pp. 108-117, 2015.
[http://dx.doi.org/10.1016/j.ijmachtools.2014.09.012]

[34] N. Haghbin, F. Ahmadzadeh, J.K. Spelt, and M. Papini, "Effect of entrained air in abrasive water jet micro-machining: Reduction of channel width and waviness using slurry entrainment", *Wear,* vol. 344-345, pp. 99-109, 2015.
[http://dx.doi.org/10.1016/j.wear.2015.10.008]

[35] R. Tripathi, M. Srivastava, S. Hloch, P. Adamčík, S. Chattopadhyaya, and A.K. Das, "Monitoring of acoustic emission during the disintegration of rock", *Procedia Eng.,* vol. 149, no. June, pp. 481-488, 2016.
[http://dx.doi.org/10.1016/j.proeng.2016.06.695]

[36] F. Yang, T. Ren, H. Wang, B. Liu, and M. Chen, "Analysis of flow field for electrochemical machining metal screw pump stator", *Int. J. Adv. Manuf. Technol.,* vol. 89, no. 5-8, pp. 1317-1326, 2017.
[http://dx.doi.org/10.1007/s00170-016-9187-y]

[37] S. Anwar, D.A. Axinte, and A.A. Becker, "Journal of materials processing technology finite element modelling of abrasive waterjet milled footprints", *J. Mater. Process. Technol.,* vol. 213, no. 2, pp. 180-193, 2013.
[http://dx.doi.org/10.1016/j.jmatprotec.2012.09.006]

[38] W. Jianming, G. Na, and G. Wenjun, "Abrasive waterjet machining simulation by SPH method", *Int. J. Adv. Manuf. Technol.,* vol. 50, no. 1-4, pp. 227-234, 2010.
[http://dx.doi.org/10.1007/s00170-010-2521-x]

[39] R. Hou, C. Huang, and H. Zhu, "Numerical simulation ultrahigh waterjet (WJ) flow field with the high-frequency velocity vibration at the nozzle inlet", *Int. J. Adv. Manuf. Technol.,* vol. 71, no. 5-8, pp. 1087-1092, 2014.
[http://dx.doi.org/10.1007/s00170-013-5493-9]

[40] D.J. Thomas, "Characterisation of aggregate notch cavity formation properties on abrasive waterjet cut surfaces", *J. Manuf. Process.,* vol. 15, no. 3, pp. 355-363, 2013.
[http://dx.doi.org/10.1016/j.jmapro.2013.02.003]

[41] V. Bhandarkar, V. Singh, and T. V. K. Gupta, "Experimental analysis and characterization of abrasive water jet machining of Inconel 718", *Mater. Today Proc,* vol. xxxx, 2019.

[42] Y. Dong, W. Liu, H. Zhang, and H. Zhang, "On-line recycling of abrasives in abrasive water jet cleaning", *Procedia CIRP,* vol. 15, pp. 278-282, 2014.
[http://dx.doi.org/10.1016/j.procir.2014.06.045]

[43] L. Huang, J. Folkes, P. Kinnell, and P.H. Shipway, "Mechanisms of damage initiation in a titanium alloy subjected to water droplet impact during ultra-high pressure plain waterjet erosion", *J. Mater. Process. Technol.,* vol. 212, no. 9, pp. 1906-1915, 2012.
[http://dx.doi.org/10.1016/j.jmatprotec.2012.04.013]

[44] D.V. Shishkin, E.S. Geskin, and B. Goldenberg, "Application of ice particles for precision cleaning of sensitive surfaces", *J. Electron. Packag. Trans. ASME,* vol. 124, no. 4, pp. 355-361, 2002.
[http://dx.doi.org/10.1115/1.1511735]

CHAPTER 5

Innovations in Abrasive Machining

Anupam Thakur[1,3,*], **Chander Prakash**[2] and **Ramakant Rana**[3]

[1] Department of Mechanical Engineering, Maharaja Agrasen University, Baddi, Himachal Pradesh, India

[2] School of Mechanical Engineering (SME), Lovely professional University, Jalandhar, India

[3] Department of Mechanical and Automation Engineering, Maharaja Agrasen Institute of Technology, Delhi, India

Abstract: Abrasive Machining has been seen making a significant advancement in recent years. Localized erosion with intensification-based technology used in manufacturing is known as Abrasive jet machining. This paper emphasizes on the recent advances being done in the various parameters and innovative techniques like thermal assistance, completely or intermediate submerge conditions on jet used for machining *etc*. Advancement in AJM is also seen in certain areas like tribology. There are certain areas like micro machining and polishing of small micro channels which also harness its advantages. It is also found that a new way of machining has been developed using elastic erosion. Other than this, certain parameters like material of jet nozzle, its design criteria, particle size, and flow rate of jet used are reviewed too. large amount of micro machining and Composite material machining advancements has been witnessed using the Abrasive jet machining.

Keywords: Abrasive jet machining, Materials, Micro machining, Process parameters.

INTRODUCTION

Abrasive jet machining is one of the non-traditional machining processes that are specifically used for the material having low shock resistance during machining process. It is also being seen that the importance of machining has gained its value over the past few years as due to various factors in the areas like machining of a design carrying various complex structures optics, medical instruments, precision instruments [1] as well improvement in the overall quality of the required product design or features [2].

---

* **Corresponding author Anupam Thakur**: Department of Mechanical Engineering, Maharaja Agrasen University, Baddi, Himachal Pradesh, India; E-mail: anupamthakur@hotmail.com

It is also seen that the AJM found its uses in precise machining having CNC conception. AJM is being diverted the micro machining due to the changes of technology diminutization thus changing the nozzle size from macro to micro scale and changing the nozzle material to sapphire. AJM has been accompanied with high pressured pumps and harder abrasive particles for machining. Verma *et al.* [5] discussed the nozzle wear that was dependent on various parameters like length of nozzle, diameter of nozzle, nozzle inlet angle, and the orifice size. It was also seen that the nozzle traverse speed played the most important role in MRR along with the abrasive flow rate up to a certain limit as it led to decrease in surface roughness and increased kerf geometry. So, there were various researches being done but some gaps were found that need to be looked on regarding the analysis of AJM, nozzle wear in AJM and the study of various parameters on the surface integrity.

There are various researches which are been discussed in this article that shows the recent advancements being done on the basis of technological advancement, the industrial application and its diversity. Review is also performed for the various material mechanisms, material removal, influence of process parameters and nozzle wear of AJM.

LITERATURE REVIEW

Abrasive Air Jet Machining

Abrasive Air Jet Machining (AAJM) can be described as the technique that works by blasting a stream of particles at a high velocity, thus causing erosion of material [6]. It is empathized to have a good material removal rate. It has been found out that on reduction of the nozzle diameter to less than 1mm it changes into a micro machining process or Micro-AAJM. The layout can be seen in the (Fig. **1**). The in-system filter and drier are used to clean the compressed air and dry it at the same time. After this the air is passed through the adjusting valve where the pressure of air is regulated and controlled. Then finally, it is passed to the blasting gun near the workpiece. The particles, after being used for erosion along with chips, are collected with the help of dust catcher. During the machining, there is a little amount of fracture depending upon the material used, properties and the various parameters considered. After the erosion the particles, the gases are escaped. Micro AAJM is used for production of small passages like passes and holes on various brittle materials such as ceramics, glasses and carbide metals. It is also found that the abrasive particles are often seen embedded in the machine surface thus creating a negative relation when nonferrous metals or polymers are used. To machine such materials, the cryogenic assisted AJM is used that is discussed later.

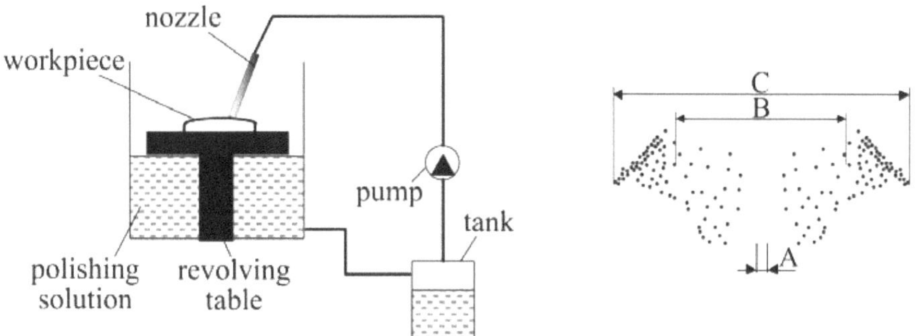

Fig. (1). Layout of Abrasive Jet Machining process [51].

Abrasive Water Jet Machining

Abrasive water jet machining is the advancement in the traditional machining process that replaces the carrier gas with the water and has found its key application in mining industry [7]. It is sometimes compared with slurry jet machining which works on two different modules of lower pressure with a premixed slurry mixture. In this, the viscosity of the medium plays an important role as the jet at the nozzle is improved up to 100 times as compared to air. It is seen that the AWJM has made a lot of improvements as roughness is reduced to nano scale [8], distant footprint [10], faster workpiece cools down and adequate slurry recovery [11, 12]. But the impact velocities are found to be different due to high deceleration in the water stagnation zone and the high influence of viscosity on trajectory of abrasive (Figs. **2** & **3**). After this the abrasive particles lose all its kinetic energy. Higher MRR is seen in air as compared to water as it provides less stagnation zone. Some research has also been done to reduce this stagnation zone [13] and proposed the usage of V shaped masks. It is also seen that use of soybean oil as carrier of abrasive can reduce the secondary impact on the workpiece in primary footprint [14]. The disadvantage of this is seen in the form of crack formation and its propagation due to the pressure of fluid used especially when working on ceramics [15].

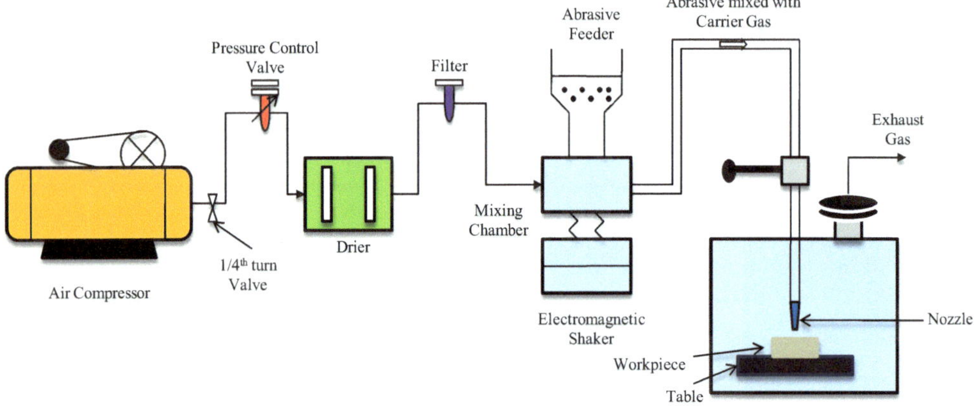

Fig. (2). Abrasive Water Jet Machining layout [3].

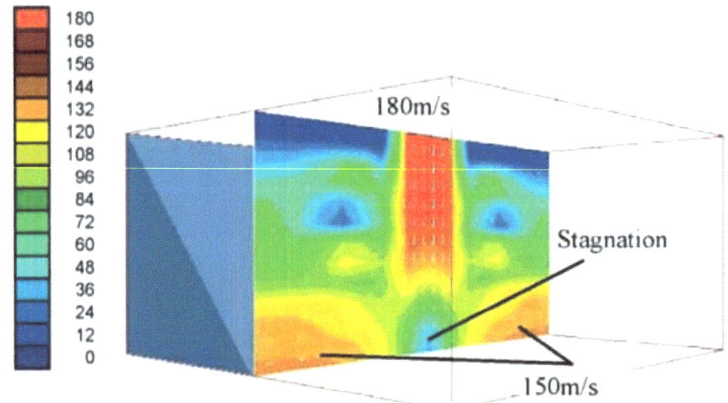

Fig. (3). Flow Velocity Distribution [9].

Use of Cryogenic in AJM

It has been discussed earlier that the softer metals cannot be easily machined having precise dimension using abrasive in AJM. Due to the small abrasives having high kinetic energy and high velocity penetrate the surface especially when hitting normal to the surface. Thus, it is found that a large amount of material cannot be machined having ductile or soft material properties by the same principle as of brittle or hard material erosion is carried out. Thus, to work on such materials cryogenics has been introduced so that the material removal mechanism can somehow be changed. AA5083-H32 aluminium alloy was machined using the both AAJM and cryogenic assisted AJM. A comparison study was carried out for the surface integrity [16]. In this study they used liquid nitrogen at -195°C and supplied it to the machining zone in compressed air. This led to the change in the hardness to a very high amount thus reducing the number of abrasives embedded in the material airing erosion. It was accompanied by the crest and troughs profile having continuous roughness. The liquid Nitrogen used also produced the strain properties and thus hardening the material being machined as an effect of grain refinement accompanying with increased micro hardness up to two times. Some other studies were also carried out that used a heat exchanger accompanied with liquid nitrogen for cooling the air stream other than the material to be machined [17, 18]. A LN2 type of heat exchanger having tubular shape carrying high and low heat thermal conductivity inside and outside respectively were used. The abrasives were directed to flow to the material at a temperature of -80°C jet condition. The temperature was thus varied from -80°C to 180°C. A change was observed when the jet temperature reached -180°C as the erosion principle changed to brittle one.

It was also accompanied in this study that the incidence angle of 30° to 60° was found to be best for maximum MRR. This study found that the erosion of softer materials can also be obtained at -180°C. To obtain more precise machining of such material with brittle removal mechanism at lower temperature can also be taken into account. It also had some disadvantages accompanied as the liquid nitrogen was being used that in its own means the sources and energy consumption on a very large scale, also such high cooling leads to crack formation due to the thermal strain in the material. But it was also found this also leads to the surface weakening which also helps in the subsequent erosion of the material.

Thermally Enhanced AJM

Patel *et al.* [19] found that there were some materials like Inconel 718 and Titanium (Ti6Al4V) that were not easily machinable; such materials were needed to have low hardness so that the abrasives can penetrate the surface thus causing the erosion to take place. For this there was some thermal assistance provided by heating the workpiece up to a temperature of 850°C by using an oxy Acetylene gas flame, thus resulting in lower hardness on the surface and increased MRR using abrasive water jet machining. Though there was a drastic change in the temperature during the process, but no stresses and crack were discovered in the machined part.

Effect of Various Jet Conditions

Intermittent Jet Conditions

It was found that when a larger grooves or cavities were made then there was an abrasive settling tendency at the bottom of such features thus preventing the upcoming new abrasive particles to reach the bottom hidden by such abrasives thereby subsequently lowering the MRR. This problem of particle accumulation was solved by creating an intermittent pattern of blocking the abrasive flow and passing a high velocity pressurized jet for that duration. This was achieved by installing a 100 Hz Solenoid valve which helped in creating this alternate pattern of abrasive and high pressure jet of air [20]. This helped in removing the accumulated abrasives at the bottom. It was found in the research that for a sample shot study of 500, the hole depth was increased from 440µm to 740 µm with an open/closing time of 10/40 milli seconds.

Submerged Jet Conditions

With the advancement of the AJM, large amount of technical and economical advantages came, however its environmental aspects also came into effect like noise created during the process. To look into this aspect a submerged AWJM was taken into consideration [21]. This somehow resulted in the lowering of the kinetic energy of the particles before reaching the machining surface. This also resulted in the reduction in the drag force on the jet thus leading to less diffraction of the jet along the diameter of nozzle [22]. This also gave the idea of having less dust in the environment during the machining as was seen in the conventional methods there by leading to some medical effects like lungs damage. It was found that a reduction of up to 61% was obtained when using the glycerin as liquid for submerging accompanied with 42% with liquid polymer and 36% with water [23]. There was a considerable increase in the nozzle life when the liquid was mixed with abrasive slurry and only jet of high velocity was directed on to the workpiece. The only drawback was the substantially low velocity of the particle.

APPLICATIONS

Abrasive jet machining has found a lot of its applications with some advantages accompanied with some drawbacks. This section discusses about some of those advantages and also some drawbacks of it. Also, its application in the areas like tribology is analyzed. Some merits are enlisted blow:

The study shows the advancement of the AJM from macro to micro scale thus giving it a boundless application in various areas and industry. Also, its large application has been seen to machine

- It has found not only its technical aspects but also its environmental aspects that could lead to better machining ability and environment possibilities thus leading to the better process flexibility.
- There has been a lot of improvements in the surface properties like grain refinement, hardening of the material as an effect of plastic deformation during

blasting [25].

- It has been found that the micro machining in occurs almost minimal cost as compared to conventional machining methods and also the abrasives can be recycled [26].

Despite all these advantages some drawbacks are seen like the tapered geometry formation while machining a groove of long depth [28], micro crack formation on a substantial brittle material, MRR is not stable in micro machining as fine particles with stable flow rate is not attained [27], rate of production in micro application is also substantially low and some critical noise levels above 105db are also observed [21].

Surface Texturing and Aspects for Tribology

Costa *et al.* found that surface texturing is a method of creating a uniform micro texture consisting of sharp impressions with controlled geometry [29]. It was found by various studies that the micro texturing gained its value as the world started to move towards advancements in electronics. It was noted that the boundary between the normal AJM and micro AJM was the orifice design criteria in shape of its opening diameter. Various studies performed show that that concise the resolution of the channel homogeneity, up to 5% for a path of 100μm using 23μm size abrasive particles and, limited it by the size the mask and the wear found at the edges [30]. It has also been summarized from the studies that the major factors limiting this are the mask material used, the fine particles size of as low as 3μm leading to its consistent flow problem along with low kinetic energy for impact and the least crack length in the workpiece material.

After attaining the capability of surface texturing, it has found its path in tribology also as shown in this section. It has been seen that in medical field a large amount of knee and hip operations and replacement have taken place all over the world. With time 80% of these have reported wear and tear within some year of use [31]. This is somewhat relating to the surface integrity and the hardness of the material being used. Out of all the sliding pair developed for artificial joint the most successful is the combination of metal to polyethylene having a 90% success rate with life of 15 years. Further studies signify the wear and tear due to surface imperfections, cracks, burrs *etc.* which leads to less joint life. It is being stated that there should be an artificial joint enhancement that leads to the collection of all the particles being produced due to wear and tear while usage and at the same time provide lubrication for its smooth functioning thus increasing it serving life up to 30 years. Such minor pores, for the collection of debris and lubrication, were difficult to machine by conventional machining methods, so, AJM has been found useful in drilling these holes. A technique to polish Cobalt-Chromiu-

-Molybdenum alloy along with engraving grooves on micro scale has been developed that could be applied in prosthesis so that least wear could occur [32]. Further other studies have been carries out that leads to much smaller holes and groves of up to 100µm and 40µm respectively. This will overall relate to the increase of the artificial knee and hip joints thus creating a large exploration area in tribology.

MATERIAL REMOVAL MECHANISM

In the earlier sections, it is witnessed that AJM has found its various applications across various materials with ductile and brittle materials along with various process parameters. Along with this, a recently found type of erosion is also seen which is named as 'elastic erosion'. Overall, the material removal mechanism (MRM) depicts the machine efficiency. As in brittle material, abrasive strikes the machine surface and creates a crack normal to the surface in downward direction, thus leading to an oblique crack at the bottom of the deformed material section area extending parallel to the top side. Here the chipped part is assumed to be hemispherical having depth or radius equal to crack created by single abrasive strike. Various studies have been done correlating the radius of chip produced depending on the toughness, hardness and viscoelasticity of the material, geometry of chip produced and force of abrasive indent for ceramics with the standard model. Authors also stated that there has to be a specific energy for a specific material for the crack propagation.

ELASTIC MODE MRM

It has been discussed earlier that the surface integrity is affected by the particle size [4]. The brittle/ductile erosion leads to the change in the surface structure at microscopic level. No surface roughness effect is found as we change the abrasive size from 120 nm to 40 nm. Thus, using particles can help of reducing the deformation and crack in final machine surface. It was thus emphasized to change the indentation mechanism to "surface area mechanism" while using abrasives of size less than 100 nm. Conventional larger sized abrasives move in a straight line on the surface, but the fine abrasive can change the direction with the flow. For a particular ratio of abrasive mass to the flow velocity, a tangential force comes into play, thus changing the path to a curved trajectory and strikes the workpiece substrate asymptotically (Fig. **4**). So, in cases where less energy for erosion is creating erosion on the substrate surface than the threshold energy, elastic erosion is said to occur. Taking this into account a new theory based on creating chemical erosion (namely elastic erosion) that considered the hydroxylation effect with chemical absorption. In this, the nano scale particles chemically react with the substrate surface to form bond within 1nm distance from substrate top layer. Peng *et al.* [33] also stated that the surface can be smoothed up to 0.17nm roughness in Ra. During this mechanism no defects were seen on the final workpiece surface.

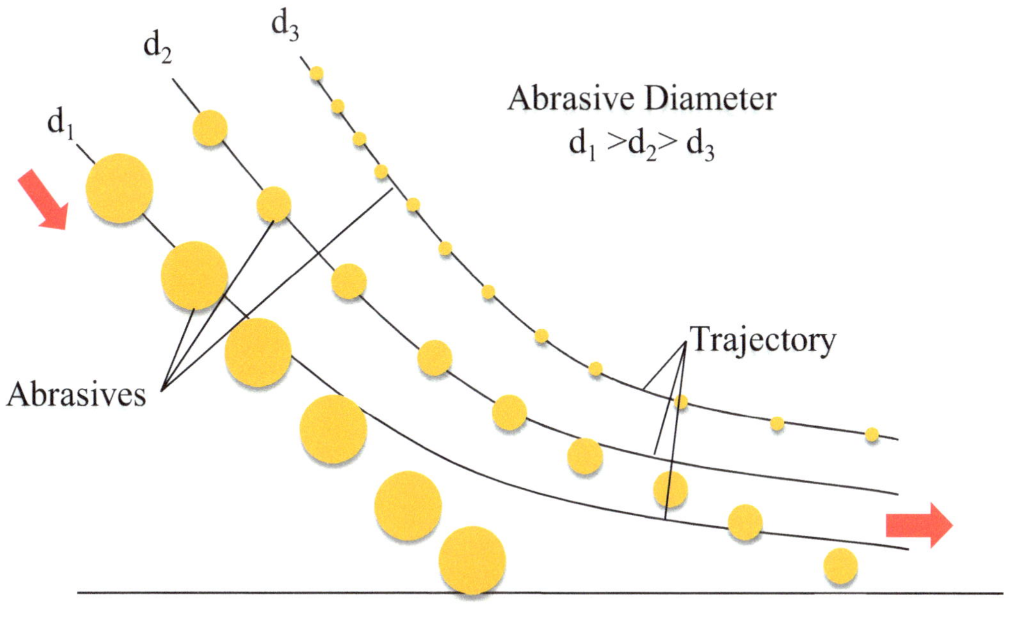

Fig. (4). Path comparison followed by different size abrasives [33].

EFFECT OF VARIOUS PROCESS PARAMETERS

Abrasive Size

Abrasive size plays a very important role in any abrasive jet machine. On increasing the size of abrasive, the volume and mass associated with it changes, in turn affecting the kinetic energy of the abrasive which affects MRR of AJM Table **1**. But the nozzle size put a limit on the abrasive size to be used even if the particles are 15 times smaller than the size of nozzle. It is found that this increase in particles size also results in the surface roughness and larger chips produced [34]. Kowsari *et al.* [35] stated in their study that the particles with size of 25μm and 10μm produced similar crest to trough roughness on working of sintered ceramic material. In case of ceramics the chips were removed by the detachment of grains from the substrate and then combining with the metal grain with more binding strength. Another factor that was seen was the impingement of abrasive particle on the workpiece surface that was believed to be due to the static friction between the two. It was also found that the fine particles tend to have lower tendency to embed as compared to larger abrasive size [36].

Flow Rate

Flow rate is defined as the gram of abrasive particles being supplied in a respective quantity of carrier gas, liquid or slurry. Li *et al.* [37] showed that the flow dynamics can be also dependent on other process factors than this. It was shown that the shape of groove produced changed to somewhat taper as the flow rate was reduced. Use of dessicant, maintaining low level of abrasive in reservoir *etc.* were stated to have best abrasive flow.

Nozzle Material and Geometry

Like the entire machining processes tool wear in the form of nozzle wear occurs across all types of AJM.

It is the most delicate part of an AJM. A high-pressure carrying gas of medium having particles is passed through the nozzle orifice. During the passage, there is collision of particles many times on the wall of the nozzle and thus leading to its wear and its parts profile disturbance, process instability, jet disturbance. Thus, it is very important to consider this as the area of improvement. The nozzle wear is measured in terms of 'nozzle weight loss rate' (NWLR) in grams/minute. Out of a various proposed nozzle material, alumina ceramic nozzle serves a greater life as compared to stainless steel nozzle. 30% less NWLR was found in monolithic B_4C coating compared to Al_2O_3/Tungsten, Titanium coating. Along with this functionally graded materials (FGM) were being found whose coating proved to be more wear resistant up to 4 times of standard nozzles. Lower stress levels at the nozzle entrance were also seen by simulating an FEM model of the same. At the same time increasing the length of the nozzle decreases the wear rate at the exit. Doing this, the particle velocity vector attained the same direction along the nozzle length thus reducing the wear. Increasing the nozzle length from 76mm to 126 mm, the life of nozzle doubled. Optimum value for Orifice: Nozzle diameter found to be 0.4 [38]. It was found that large amount of nozzle wear also take place due to the tensile and thermal stresses. The optimum angle for nozzle entrance was discovered to be 11°. At the same time, the Larger entrance angles produced more stresses.

Table 1. Review of Material, parameters and objective considered in AWJM.

S.No.	Author	Abrasive Material	Work Material	Influential Parameters	Result
1	Paul *et al.* [39]	Garnet	C-37 Steel	Orifice dia.- 0.3 mm, Inclination angle- 90°, Insert diameter 1.1 mm, Water Pressure- 2000-3000 bar, Abrasive Flow rate-0.22--0.832 Kb/min, Traverse speed- 2000-4000mm/min, standoff dist.- 4-10mm, Feed-0.25-0.8	Traverse speed, Stand off distance and traverse feed were found to be most influential with Ra of 4-6μ.
2	AM Hoogstrate *et al.* [40]	Garnet Olivine	Glass Fibre metal laminates	Orifice diameter-0.3 mm, Insert Diameter of 1.1 mm, length 51 mm, Impact angle 90°, Abrasive size Mesh 60 Water pressure -2000 to 3000 bar, Abrasive Mass Flowrate 0.145 to 0.92 kg/min, Stand-off distance - 2 mm, Traverse speed 500 to 3000mm/min	The laminated layer was most influenced by the water pressure. The cutting ability parameter (C_p) was in direct relation by the kerf quality parameters.
3	Hussein *et al.* [41]	Al_2O_3	GRPF Composite type 3240	Hole diameter- 6 to 8mm, Cutting Speed- 0.2to 0.3 m/min, Abrasive flow rate – 100 to 130 gm/min. Jet pressure – 150 MPa, Standoff distance – 2mm.	Cutting feed, Stand off dist. and jet pressure have most influence on the surface roughness. Ra of 0.402 μm was obtained during machining.
4	Haque *et al.* [42]	Garnet, Al_2O_3 and SiC	Glass	Orifice Dia- 0.1mm, Jet orientation 90°, SOD- 1-5mm, Feed rate -10-50 mm/min., Jet pressure- 10 to 50 ksi,	Taper of cut increased with SOD increase, Garnet produces more taper as compared to other two. Direct relation between the taper and work feed rate. Avg. cut and cutting ability is higher for SiC than Al_2O_3 and garnet.
5	Vu Ngoc Pi *et al.* [43]	GMA abrasive (New and Recycled)	Al6061T6	Orifice diameter-0.25 mm, Nozzle Dia-0.92mm, Impact angle 90°, Abrasive size 63 to 180 μ, Water pressure -300 to 360MPa, Abrasive Mass Flowrate 50 to 300g/min, Stand-off distance - 2 mm, Traverse speed 120mm/min.	Optimal size for new and recycled abrasive was 90μm and better cutting performance using recycled abrasives with surface roughness of 6μm.

(Table 1) cont.....

S.No.	Author	Abrasive Material	Work Material	Influential Parameters	Result
6	Khajavi *et al.* [44]	Garnet	AA6063-T6	Water pressure-165 to 239 MPa, Traverse speed 37 to 125 mm/min, Abrasive flow rate - 0.35 to 0.95 kg/min., and diameter of focusing nozzle- 0.8 to 1.5 mm.	Model was developed for predicting the process parameters with output characteristics using ANOVA within 1% error.
7	Gulhane *et al.* [45]	Black emery particles	Ceramic	Nozzle Diameter 3 to 5 (mm) , Pressure 4-6 (kgf/cm , Stand of Distance 20 to 40(mm)	Results are matched of Taguchi approach with ANOVA and the effect of nozzle dia is most seen on the MRR and SOD is most effecting the kerf width
8	Rao *et al.* [46]	Garnet (36% FeO, 33% SiO_2, 20% Al_2O_3, 4%MgO, 3%TiO_2, 2%CaO and 2%MnO_2).	Mild Steel	Water pressure 35000 -55000 psi, Traverse speed 80-320 m/min., SOD – 1to 4 mm, Jet impact angle- 90°,nozzle dia- 0.35 mm, nozzle length – 76.2mm, nozzle dia.-1.05mm, dia of abrasive 0.18 mm	Traverse speed has most influence on Surface roughness as compared to water pressure and SOD. Optimal parameters for least SR are Water pressure – 35000 psi, Speed – 80mm/min an SOD 2.5mm
9	S.Srinivas al [47].	Garnet	Inconel 718	Water Pressure – 100 to 300 Mpa, Traverse Speed – 75-175 mm/min., Abrasive Flow rate – 0.3017 to 0.5994 kg/min, Orifice Dia 0.25 mm, Jet impact angle 90°, Focusing nozzle Diameter 1.07mm	Traverse Speed and water pressure are most important factors for depth of penetration.
10	Santanakumar *et al.* [48]	Garnet	Ceramic Tiles	Abrasive Grain Size - #80 to 120, Abrasive Flow Rate - 300 to 500 gm/min., SOD - 1 to 3 mm, Water Pressure-2000 to 3000 bar, Jet Traverse Rate - 200 to 400 mm/min.	A prediction model was propsed to optimize the Surface Roughness and Taper Angle which were found to be an avg. of 5.27 μm and 1.41°.

(Table 1) cont.....

S.No.	Author	Abrasive Material	Work Material	Influential Parameters	Result
11	Kanthababu *et al.* [49]	Garnet	Al7075 (0 to15%) with B_4C Composite	Orifice Dia – 0.25 mm, Focus Nozzle Dia 0.75mm, Jet impact angle 90°, Mesh Size - 80 to 120#, Abrasive flow Rate 240 to 440 g/min., Water Pressure- 125 to 275 MPa, Traverse rate- 60 to 120 mm/min.	DOC for each composite was measured and optimal condition were found for the cutting characteristics. DOC of upto 25 mm were obtained for metal composite of Al7075 5% with B_4C.

DOC-Depth Of cut, # = Mesh Size, SR/Ra- Surface Roughness, SOD= Standoff Distance, MRR= Material Removal Rate.

CONCLUSION

AJM is a modern technique of machining that offers machining on micro to macro scale having much higher precision and control. Various types of AJM are being discussed with their benefits and limitations in terms of accuracy, sustainability of better environment and larger sized machining *etc.* It is being studied that the tribological applications of AJM seem to be of quite relevant and more research should be done to enhance this to other fields. Introduction of cryogenics for machining of softer material to offer precise machining was also studied. It came with other drawbacks like thermal stresses which should be studied. This also need to be minimized as the temperature and thermal effect play a very important role in achieving higher effect on machining efficiency. Research could also be done on the incident angle for AJM as it comes out to be an important factor in Abrasive fluid jet polishing (AFJP) [50]. The wear characteristics of the jet nozzle are also not studied in a thorough manner thus it could also be done. Though FGM provide much improvement in the nozzle wear but more materials like micro sapphire and diamond nozzles also could be studied to check their performance characteristics.

CONSENT FOR PUBLICATION

Not applicable.

CONFLICT OF INTEREST

The authors confirm that this chapter content has no conflict of interest.

ACKNOWLEDGEMENTS

The authors would like to express their sincere thanks to the editor and anonymous reviewers for their time and valuable suggestions.

REFERENCES

[1]   F.J. Chen, S.H. Yin, H. Ohmori, and J.W. Yu, "Form error compensation in single-point inclined axis nanogrinding for small aspheric insert", *Int. J. Adv. Manuf. Technol.,* vol. 65, no. 1–4, pp. 433-441, 2013.
        [http://dx.doi.org/10.1007/s00170-012-4182-4]

[2]   F.J. Chen, S.J. Hu, and S.H. Yin, "A novel mathematical model for grinding ball-end milling cutter with equal rake and clearance angle", *Int. J. Adv. Manuf. Technol.,* vol. 63, no. 1–4, pp. 109-116, 2012.
        [http://dx.doi.org/10.1007/s00170-011-3889-y]

[3]   A. Ray, and S. Jagadish, *Bhowmik, "Abrasive water jet machining of composite materials.* Advanced Manufacturing Technology, 2017, pp. 77-97.

[4]   H. Fang, P.J. Guo, and J.C. Yu, "Research on material removal mechanism of fluid jet polishing", *Optimization Technique,* vol. 30, no. 2, pp. 248-250, 2004.

[5]   S. Verma, S.K. Moulick, and S.K. Mishra, "Nozzle Wear Parameter in Water Jet Machining The Review", *International Journal of Engineering Development and Research,* vol. 2, pp. 1063-1073, 2014.

[6]   M. Wakuda, Y. Yamauchi, and S. Kanzaki, "Effect of Workpiece Properties on Machinability in Abrasive Jet Machining of Ceramic Materials", *Precis. Eng.,* vol. 26, pp. 193-198, 2002.
        [http://dx.doi.org/10.1016/S0141-6359(01)00114-3]

[7]   M. Molitoris, J. Pitel, A. Hosovsky, M. Tothova, and K. Zidek, "A Review of Research on Water Jet with Slurry Injection", *Procedia Eng.,* vol. 149, pp. 333-339, 2016.
        [http://dx.doi.org/10.1016/j.proeng.2016.06.675]

[8]   Z.Z. Li, J.M. Wang, X.Q. Peng, L.T. Ho, and Z.Q. Yin, "S.Y. Li,C.F. Cheung, "Removal of Single Point Diamond-turning Marks by Abrasive Jet Polishing", *Appl. Opt.,* vol. 50, pp. 24-58, 2011.
        [http://dx.doi.org/10.1364/AO.50.002458]

[9]   T. Matsumura, T. Muramatsu, and S. Fueki, *Abrasive Water Jet Machining of Glass with Stagnation Effect.* vol. Vol. 60. CIRP Annals – Manufacturing Technology, 2011, pp. 355-358.

[10]   N. Haghbin, J.K. Spelt, and M. Papini, "Abrasive Waterjet Micro-machining of Channels in Metals: Comparison Between Machining in Air and Submerged in Water", *Int. J. Mach. Tools Manuf.,* vol. 88, pp. 108-117, 2015.
        [http://dx.doi.org/10.1016/j.ijmachtools.2014.09.012]

[11]   H.C. Li, and W.S. Chen, "Recovery of Silicon Carbide from Waste Silicon Slurry by Using Flotation", *Energy Procedia,* vol. 136, pp. 53-59, 2017.
        [http://dx.doi.org/10.1016/j.egypro.2017.10.281]

[12]   J.Y. Kim, U.S. Kim, M.S. Byeon, W.K. Kang, K.T. Hwang, and W.S. Cho, "Recovery of Cerium from Glass Polishing Slurry", *J. Rare Earths,* vol. 29, pp. 1075-1078, 2011.
        [http://dx.doi.org/10.1016/S1002-0721(10)60601-1]

[13]   T. Matsumura, T. Muramatsu, and S. Fueki, *Abrasive Water Jet Machining of Glass with Stagnation Effect.* vol. Vol. 60. CIRP Annals – Manufacturing Technology, 2011, pp. 355-358.

[14]   K. Kowsari, M. Papini, and J.K. Spelt, "Selective Removal of Metallic Layers from Sintered Ceramic and Metallic Plates Using Abrasive Slurry-jet Micro-machining", *J. Manuf. Process.,* vol. 29, pp. 252-264, 2017.
        [http://dx.doi.org/10.1016/j.jmapro.2017.08.005]

[15]   S. Lathabai, and D.C. Pender, "Microstructural Influence in Slurry Erosion of Ceramics", *Wear,* vol. 189, pp. 122-135, 1995.
[http://dx.doi.org/10.1016/0043-1648(95)06679-9]

[16]   N. Yuvaraj, and M.P. Kumar, "Cutting of Aluminium Alloy with Abrasive Water Jet and Cryogenic Assisted Abrasive Water Jet: A Comparative Study of the Surface Integrity Approach", *Wear,* vol. 362–363, pp. 18-32, 2016.
[http://dx.doi.org/10.1016/j.wear.2016.05.008]

[17]   A.G. Gradeen, J.K. Spelt, and M. Papini, "Cryogenic Abrasive Jet Machining of Polydimethylsiloxane at Different Temperatures", *Wear,* vol. 274–275, pp. 335-344, 2012.
[http://dx.doi.org/10.1016/j.wear.2011.09.013]

[18]   A.G. Gradeen, M. Papini, and J.K. Spelt, "The Effect of Temperature on the Cryogenic Abrasive Jet Micro-machining of Polytetrafluoroethylene, High Carbon Steel and Polydimethylsiloxane", *Wear,* vol. 317, pp. 170-178, 2014.
[http://dx.doi.org/10.1016/j.wear.2014.06.002]

[19]   D. Patel, and P. Tandon, "Experimental Investigations of Thermally En-hanced Abrasive Water Jet Machining of Hard-to-machine Metals", *CIRO J. Manuf. Sci. Technol.,* vol. 10, pp. 92-101, 2015.
[http://dx.doi.org/10.1016/j.cirpj.2015.04.002]

[20]   L. Zhang, T. Kuriyagawa, Y. Yasutomi, and J. Zhao, "Investigation into Micro Abrasive Intermittent Jet Machining", *Int. J. Mach. Tools Manuf.,* vol. 55, pp. 873-879, 2005.
[http://dx.doi.org/10.1016/j.ijmachtools.2004.11.003]

[21]   A. Radvanska, T. Ergic, Z. Ivandic, S. Hloch, J. Valicek, and J. Mullerova, "Technical Possibilities of Noise Reduction in Material Cutting by Abrasive Water-jet", *Strojarstvo,* vol. 51, pp. 347-354, 2009.

[22]   N. Haghbin, J.K. Spelt, and M. Papini, "Abrasive Waterjet Micro-machining of Channels in Metals: Model to Predict High Aspect-ratio Channel Profiles for Submerged and Unsubmerged Machining", *J. Mater. Process. Technol.,* vol. 222, pp. 399-409, 2015.
[http://dx.doi.org/10.1016/j.jmatprotec.2015.03.026]

[23]   R.H.M. Jafar, V. Hadavi, J.K. Spelt, and M. Papini, "Dust Reduction in Abrasive Jet Micro-machining Using Liquid Films", *Powder Technol.,* vol. 301, pp. 1270-1274, 2016.
[http://dx.doi.org/10.1016/j.powtec.2016.08.002]

[24]   C.J. Wang, C.F. Cheung, L.T. Ho, M.Y. Liu, and W.B. Lee, "A Novel Multi-jet Polishing Process and Tool for High-efficiency Polishing", *Int. J. Mach. Tools Manuf.,* vol. 115, pp. 60-73, 2017.
[http://dx.doi.org/10.1016/j.ijmachtools.2016.12.006]

[25]   S. Barriuso, M. Jaafar, J. Chao, A. Asenjo, and J.L. Gonzalez-Carrasco, "Improvement of the Blasting Induced Effects on Medical 316 Lvm Stainless Steel by Short-term Thermal Treatments", *Surf. Coat. Tech.,* vol. 258, pp. 1075-1081, 2014.
[http://dx.doi.org/10.1016/j.surfcoat.2014.07.027]

[26]   Y. Dong, W. Liu, H. Zhang, and H. Zhang, "On-line Recycling of Abrasives in Abrasive Water Jet Cleaning", *Procedia CIRP,* vol. 15, pp. 278-282, 2014.
[http://dx.doi.org/10.1016/j.procir.2014.06.045]

[27]   R. Shukla, and D. Singh, "Experimentation Investigation of Abrasive Water Jet Machining Parameters Using Taguchi and Evolutionary Optimization Techniques", *Swarm Evol. Comput.,* vol. 32, pp. 167-183, 2017.
[http://dx.doi.org/10.1016/j.swevo.2016.07.002]

[28]   N.S. Pawar, R.R. Lakhe, and R.L. Shrivastava, "Validation of Experimental Work by Using Cubic Polynomial Models For Sea Sand as an Abrasive Material in Silicon Nozzle in Abrasive Jet Machining Process", *Materials Today: Proceedings,* vol. 2, pp. 1927-1933, 2015.

[29]   H. Costa, and I. Hutchings, "Some Innovative Surface Texturing Techniques for Tribological Purposes", *Proc. Inst. Mech. Eng., Part J J. Eng. Tribol.,* vol. 229, pp. 429-448, 2015.

[http://dx.doi.org/10.1177/1350650114539936]

[30] P.J. Slikkerveer, P.C.P. Bouten, and F.C.M. De Haas, "High Quality Mechanical Etching of Brittle Materials by Powder Blasting", *Sens. Actuators A Phys.,* vol. 85, pp. 296-303, 2000.
[http://dx.doi.org/10.1016/S0924-4247(00)00343-5]

[31] L. Kuncicka, R. Kocich, and T.C. Lowe, "Advances in Metals and Alloys for Joint Replacement", *Prog. Mater. Sci.,* vol. 88, pp. 232-280, 2017.
[http://dx.doi.org/10.1016/j.pmatsci.2017.04.002]

[32] Y. Nakanishi, Y. Nakashima, Y. Fujiwara, Y. Komohara, M. Takeya, H. Miura, and H. Higaki, "Influence of Surface Profile of Co-28Cr-6Mo Alloy on Wear Behaviour of Ultra-high Molecular Weight Polyethylene Used in Artificial Joint", *Tribol. Int.,* vol. 118, pp. 538-546, 2018.
[http://dx.doi.org/10.1016/j.triboint.2017.06.030]

[33] W. Peng, C. Guan, and S. Li, "Material removal mode affected by the particle size in fluid jet polishing", *Appl. Opt.,* vol. 52, no. 33, pp. 7927-7933, 2013.
[http://dx.doi.org/10.1364/AO.52.007927] [PMID: 24513743]

[34] N. Su, L. Yue, Y. Liao, W. Liu, H. Zhang, X. Li, H. Wang, and J. Shen, "The effect of various sandblasting conditions on surface changes of dental zirconia and shear bond strength between zirconia core and indirect composite resin", *J. Adv. Prosthodont.,* vol. 7, no. 3, pp. 214-223, 2015.
[http://dx.doi.org/10.4047/jap.2015.7.3.214] [PMID: 26140173]

[35] K. Kowsari, M.R. Sookhaklari, H. Nouraei, M. Papini, and J.K. Spelt, "Hybrid Erosive Jet Micro-milling of Sintered Ceramic Wafers with and without Copper-filled Through-holes", *J. Mater. Process. Technol.,* vol. 230, pp. 198-210, 2016.
[http://dx.doi.org/10.1016/j.jmatprotec.2015.11.027]

[36] V. Hadavi, B. Michaelsen, and M. Papini, "Measurements and Modeling of Instantaneous Particle Orientation within Abrasive Air Jets and Implications for Particle Embedding", *Wear,* vol. 336–337, pp. 9-20, 2015.
[http://dx.doi.org/10.1016/j.wear.2015.04.016]

[37] H. Li, J. Wang, N. Kwok, T. Nguyen, and G.H. Yeoh, "A Study of the Micro-hole Geometry Evolution on Glass by Abrasive Air-jet Micromachining", *J. Manuf. Process.,* vol. 31, pp. 156-16, 2018.
[http://dx.doi.org/10.1016/j.jmapro.2017.11.013]

[38] M. Nanduri, D.G. Taggart, and T.J. Kim, "The Effects of System and Geometric Parameters on Abrasive Water Jet Nozzle Wear", *Int. J. Mach. Tools Manuf.,* vol. 42, pp. 615-623, 2002.
[http://dx.doi.org/10.1016/S0890-6955(01)00147-X]

[39] S. Paul, A.M. Hoogstrate, C.A. van Luttervelt, and H.J.J. Kals, "An experimental investigation of rectangular pocket milling with abrasive water jet", *J. Mater. Process. Technol.,* vol. 73, pp. 179-188, 1998.
[http://dx.doi.org/10.1016/S0924-0136(97)00227-6]

[40] S. Paul, A.M. Hoogstrate, and R. van Praag, "Abrasive water jet machining of glass fibre metal laminates", *Proceedings of the Institution of Mechanical,* vol. 216, no. issue: 11, pp. 1459-1469, 2002.
[http://dx.doi.org/10.1243/095440502320783396]

[41] M.A.I. Hussein, A. Iqbalb, and M. Hashemipourb, "Numerical optimization of hole making in GFRP composite using abrasive water jet machining process", *Zhongguo Gongcheng Xuekan,* vol. 38, pp. 66-76, 2015.
[http://dx.doi.org/10.1080/02533839.2014.953240]

[42] A.A. Khan, and M.M. Haque, "Performance of different abrasive materials during abrasive water jet machining of glass", *J. Mater. Process. Technol.,* vol. 191, pp. 404-407, 2007.
[http://dx.doi.org/10.1016/j.jmatprotec.2007.03.071]

[43] V.N. Pi, and A.M. Hoogstrate, "A study on abrasive recycling and recharging in abrasive water jet

(AWJ) machining", *Int. J. Machining and Machinability of Materials,* vol. 6, no. 3/4, pp. 213-233, 2009.
[http://dx.doi.org/10.1504/IJMMM.2009.027325]

[44]   A. Farhad Kolahan, "A statistical approach for predicting and optimizing depth of cut in AWJ machining for 6063-T6 Al alloy", *World Academy of Science, Engineering and Technology International Journal of Mechanical and Mechatronics Engineering,* vol. 3, no. 11, 2009.

[45]   U.D. Gulhane, P.P. Patkar, P.P. Toraskar, S.P. Patil, and A.A. Patil, "Analysis of abrasive jet machining parameters on mrr and kerf width of hard and brittle materials like ceramic", *International Journal of Design and Manufacturing Technology,* vol. 4, no. 1, pp. 51-58, 2013. [IJDMT].

[46]   M. Sreenivasa Rao, S. Ravinder, and A. Seshu Kumar, "Parametric Optimization of Abrasive Waterjet Machining for Mild Steel: Taguchi Approach", *International Journal of Current Engineering and Technology, Special Issue,* vol. 2, pp. 28-30, 2014.

[47]   Raghavendra , Y. Ratnagiri, and S. Srinivas, "Effect of Process Parameters on Abrasive Water Jet Penetration Capabilities on Inconel 718", *International Conference on Precision (COPEN 10),* pp. 404-407, 2017.

[48]   M. Santhanakumar, R. Adalarasan, and M. Rajmohan, "Experimental Modelling and Analysis in AbrasiveWaterjet Cutting of Ceramic Tiles Using Grey-Based Response Surface Methodology", *Arab. J. Sci. Eng.,* vol. 40, pp. 3299-3311, 2015.
[http://dx.doi.org/10.1007/s13369-015-1775-x]

[49]   V. Mohankumar, and M. Kanthababu, "Experimental Investigations on Depth of Cut in Abrasive Waterjet Machining of Al7075/B4C Metal Matrix Composites", *Int. J. Engine Res.,* vol. 5, no. 4, pp. 322-326, 2016.

[50]   C.J. Wang, C.F. Cheung, and M.Y. Liu, "Numerical Modeling and Experimentation of Three Dimensional Material Removal Characteristics in Fluid Jet Polishing", *Int. J. Mech. Sci.,* vol. 133, pp. 568-577, 2017.
[http://dx.doi.org/10.1016/j.ijmecsci.2017.09.018]

CHAPTER 6

A Review on Electrical Discharge Machining of Ceramic Based Composite Material

Mamta [1,*]**, Suneev Anil Bansal**[1] **and Bhuvnesh Kumar**[2]

[1] *Department of Mechanical Engineering, Maharaja Agrasen University, Baddi, Solan, India*

[2] *Mechanical Engineering Department, SRM University Delhi NCR Haryana, India*

Abstract: EDM is the non-conventional process that has used, extensively, for cutting harder like nickel and its alloys, titanium and complex shape materials which are difficult to machine by other conventional processes. EDM utilizes, electrical and heat, energy to machine the workpiece by removing the undesirable material. This process use two electrodes, anode and cathode out of which tool act as cathode and the workpiece on which the job is being performed act as anode. Various factors like current, pulse on and off time, voltage applied can be used to optimize the major machining parameters of EDM like MRR, rate of tool wear, surface roughness *etc.* the main focus of this study these parameters on composite materials made of CMs, tungsten and cobalt carbides *etc.* In the end future prospective of the areas which require further improvements and investigations has also been discussed.

Keywords: Ceramic, Composite materials, Electrical discharge machining.

INTRODUCTION

Electro discharge machining (EDM) is a prevailing manufacturing process that is used for the machining of metals. In this process the machining take place in the form of melting and evaporation of all materials that are electrically conductive in nature which occur by controlling the spark between the electrode and the work piece on which the operation is being performed [1, 2]. This is the reason EDM is also known as nonconventional spark eroded thermo- electric machining process [3]. Over a course of period EDM has emerged as an efficient process for removal of the materials. EDM holds the capacity of machining very complex and hard materials, that too with high precision which are otherwise difficult to machine with other conventional methods.

---

[*] **Corresponding author Mamta**: Department of Mechanical Engineering, Maharaja Agrasen University, Baddi, Solan, India; E-mail: dahiya.mamta56@gmail.com

Suneev Anil Bansal (Ed.)

To fall on the footsteps of modern manufacturing industry, EDM also possess the capability of being an automated process. EDM finds its major application in the manufacturing of prototypes, in mold and die making sectors and in the industries which are mostly indulged in the making of machined metals such as high alloyed steels.

Past couple of decades, ceramic materials (CMs) have marked their presence by their application in all major engineering sectors like automobile, airspace, sectors involving manufacturing of electrical and electronic equipment [4]. Some major applications like fabrication of the ballistic armors, braking system of automobiles, particulate filters for diesel, memory components of computers, sensors and various medical products like prosthetic body limbs *etc.* also involve the use of CMs [5]. The reason for so much wider applications of CMs are, its excellent mechanical properties such as high compressive strength, high hardness, chemical and abrasive resistance.

WC–Cos were also explored for machining by EDM. A special CIRP reported used of EDM in cemented carbide too [6]. Although, it is reported that EDM surfaces do have residual thermal stress [7]. group from

PRINCIPLE OF EDM

In EDM, the removal of the material is obtained by controlling the electrical discharges between the two electrodes out of which one electrode is the tool and the other one is the work piece on which the job is being performed as shown in Fig. (**1**).

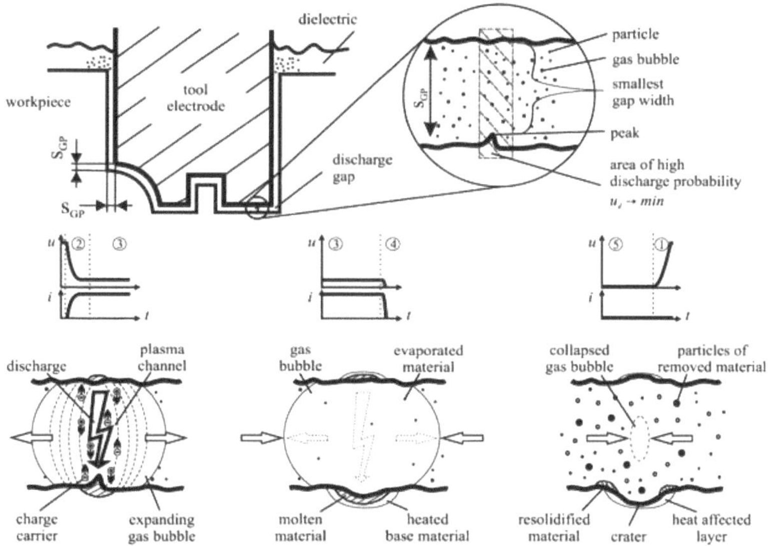

Fig. (1). Principle of EDM process [8].

Both the electrodes are separated by a certain dielectric medium which may be either dielectric oil or deionised water. Then the voltage is applied thus forming a plasma channel that can establish a flow of the current. When the breakdown in voltage medium is reached a discharge takes place. As the process of EDM is based on the removal of material by melting and evaporation therefore when the temperature at the base of the formed channel reached to approximately $\geq$ 10,000°K, the melting and evaporating of the second electrode (workpiece) takes place thus resulting in the removal of material.

ELECTRICAL DISCHARGE MACHINING OF CERAMICS

CMs, like titanium diboride (TiB_2), silicon doped silicon carbide (SiSiC) and titanium nitride (TiN), are electrically conductive in nature and can be machined by EDM [9]. According to Luis *et al.* [10] the materials which are obtained from inorganic materials, generally having the purity of high grade, are referred as ceramics or advanced CMs. These CMs are further classified into oxide ceramics like Al_2O_3 or ZrO_2 and non-oxide ceramics like Si_3N_4 or SiC.

The machining of CMs of with the process of EDM can be also classified into natural conductive materials, non-conductive materials and conductive materials. Mostly the conductive materials are made from non- conductive material by mixing some impure atoms and it is done by introducing some electro-conductive secondary phases (Fig. **2**).

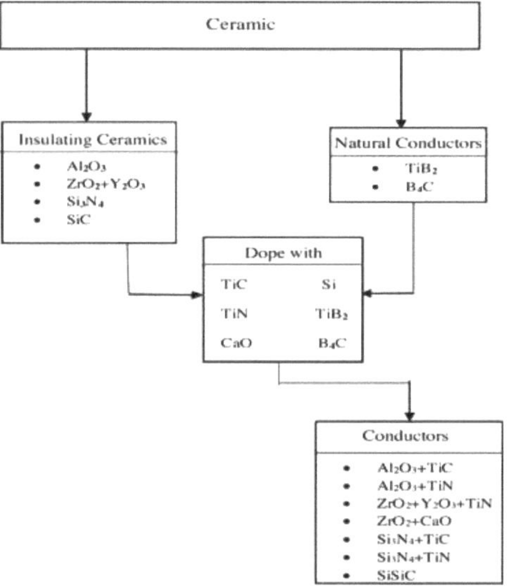

Fig. (2). Ceramic materials classification based on their Conductivity [11].

Adding these phases to the CMs thus increases the electrical conductivity. Some improvement in the hardness, strength and fracture toughness is reported too by adding these doping agents [12, 13].

MECHANISM OF MATERIAL REMOVAL IN COMPOSITE CERAMICS

Lauwers *et al.* [13] conducted a study to develop CMs which were wear resistant and also able to be machine by the process of EDM. The main focus of this study was to investigate the relation between the parameters of EDM process and the microstructure of the CMs. For this ZrO_2, Al_2O_3 and Si_3N_4 based CMs were taken and to these materials, some phases of electrical conductivity were induced by TiCn and TiN. In the results, it was observed that not only melting and evaporation but other factors like oxidation and base material decomposition are responsible for effective removal of the material. Also it was started that the formations of the crack during the process were resulted due to the spalling effect. The main mechanisms for removal of material are presented in Table **1**

Table1. Mechanism for material removal of different composite CMs [13].

Material Removal Mechanism in Different Composite Ceramics			
Composite Ceramic	Dielectric	Machine Configuration	Material Removal Mechanism
ZrO_2-TiN.	Deionized water.	WEDM (Wire EDM with Roughing condition)	Melting and Evaporation.
Al_2O_3-SiCw-TiC.	Deionized water.	WEDM (Wire EDM with Roughing condition)	Melting Evaporation.
Al_2O_3-SiCw-TiC.	Oil.	Die-Sinking EDM with Medium and high energy input.	Spalling.
Si_3N_4-TiN	Deionized water.	Die-Sinking EDM High energy input.	Spalling.
Si_3N_4-TiN.	Deionized water.	WEDM.	Oxidation. Decomposition.
Si_3N_4-TiN.	Oil.	Die-Sinking. EDM with Roughing condition	Melting. Evaporation.

It is observed from Table **1** that dielectric medium, discharge medium and configuration of machining parameters plays an important role in mechanism of the material removal. The two major mechanisms are melting/evaporation and spalling. Melting and evaporation takes place in the state of low energy discharge while the spalling mechanism takes place in high energy discharge state. Overview of the electrical conductivity of some ceramics and metallic materials is shown in Fig. (**3**). It is observed that like the metallic materials, ceramics that are electrically conductive in nature can also be easily machined by the EDM process *e.g.* titanium diboride (TiB_2), titanium nitride (TiN), or silicon doped silicon carbide (SiSiC) [14].

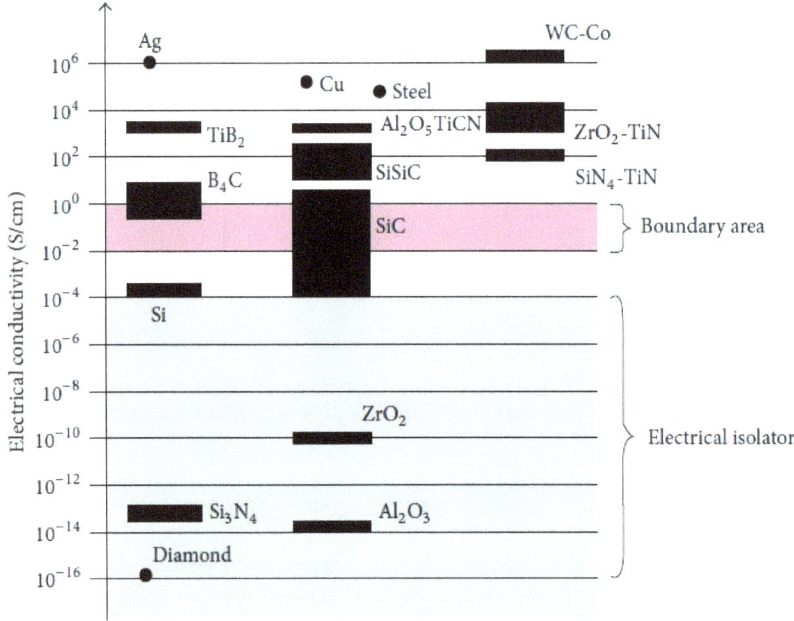

Fig. (3). Electrical conductivity of materials [15].

MACHINING OF WC-CO COMPOSITE MATERIAL

Composites made of Tungsten carbide/cobalt (WC-Co) material hold good range of mechanical properties such as good strength, high value of hardness and good ample amount of wear resistance for wider range of temperatures thus find wide application in the production of dies, cutting tools and some special purpose tools. These composite materials possess very high specific strength, thus difficult to machine by conventional machining processes [16]. Also, it involves some complex shapes therefor very high accuracy is required during machining which is not possible to achieve with the conventional machining processes [17].

Number of studies reported the machining of these materials with conventional process like using of CBN tools in CNC turning. The results obtained showed negative impacts like high cutting force, low MRR and poor finish of obtained surface [18]. On the other hand, Levy *et al.* [19] & Bonny *et al.* [20] conducted some studies and observed that the process of EDM is capable of maching of these composite materials to fabricate cemented carbide dies which involves complex shapes irespective of their hardness and strenth In other study Mahdavinejad *et al.* [2] coonducted a study to analyse the instability of EDM prcesss on machining of WC-Co composite materials. In the results it was observed that for the various compositions of WC-Co composite materials, pulse duration factor effects the stock removal rate. The removal rate tends to increase with the increase in pulse duration. It was also observed the machining instability was also increased with the increase in the pulse duration. It was stated that this instability could be reduced by reducing the percentage of Co in the formed composite material (Fig. **4**).

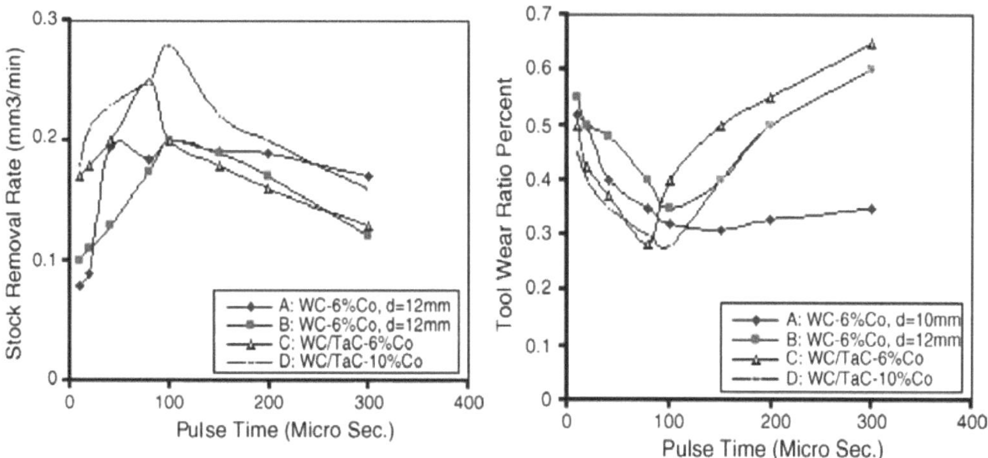

Fig. (4). Effect on different compositions of WC-Co by variation of tool wear rate, pulse time and stock removal rate [2].

Similar study was conducted and it was observed that various factors like pulse timing and pulse current greatly influence the surface roughness of the workpiece in which the job is being performed [6]. Higher the values of these machining parameters, better will be the value of surface roughness of the workpiece [6].

ELECTRICAL DISCHARGE MACHINING- LITERATURE SURVEY

Gadalla *et al.* [21] conducted experiments to examine the shaping of composites made of CMs by the process of EDM. The results were compared with the surfaces machined by other conventional processes. In the results, it was observed that the MRR was increased due to the decrease of per unit volume of energy and and the main factors like electrical resistance and thermal diffusivity were the main factors that affect the MRR. It was also stated that the surface roughness of the material largely effected by the dispersion of the grain size of the material. In the similar study Farooqui *et al.* [22] compared the machining of ceramics and advanced ceramics by EDM process. It was observed that the pulse duration of the process was responsible for the mechanism of removal of the material. While the oxidation and decomposition was due to the shorter pulse, on the other hand longer pulse was responsible for spalling mechanism. In another study Gamage *et al.* [23] used alluminium 3003 as materail with tool made of AISI P20 for conducting case studies of wire EDM and DSEDM. The results were observed and it was stated that the electrical energy was the major source of impact followed by other factors like dielectric oil and material of the wire for DSEDM and wire EDM respectively. Lauwer [24] presented a study in which overview was provided to state the various challenges related to EDM of composites made of CMs. In the results it was concluded that by altering the composition of the material both the EDM process and mechanical properties of the material can be improved to certain extent. Schubert *et al.* [15] conducted a study to present the recent studies related to micro EDM of composites made of CMs. Various examples of procedure, used for insulating ceramics and machining of the ceramics, were also discussed and presented. In the results its was observed that due to the method of assisting electrodes it is advisable to machine the insulating ceramics and therefore are regarded as suitable post processing process for the CMs. Bucciotti*et al.* [25] conducted experiment to process biomedical implant devices made of silicon and titanium nitride by the process of EDM. The mechanical, toxicological and microstructural properties were studied and it was observed these materials holds good as compared to the other oxide CMs used for making the same joints and also very complex shapes like undercuts and near to net shapes were possible to manufacture by EDM process.

Mohri *et al.* [26] conducted experiments to investigate the use of EDM process for the machining of insulating CMs by the use of assisting electrodes. The kerosene was used a working fluid and it was observed that by using the copper electrodes the machining of CMs was very easy. In another study wire-EDM finishing was reported to ZrO_2-TiN hybrid composite materials. Spark Plasma Sintering was used to synthesie blanks of ZrO_2-TiN [27]. Sano *et al.* [28] conducted experiments with EDM processes by using electrode made of polycrystalline diamond (PCD). Two kinds of PCB namely CTB-010 and CTH-015, both with the flat surfaces were used and the results were compared with the electrodes made of Cu-W. It was observed that the electrodes made with PCD shows less wear of 1.5% for short pulse duration and no wear for long pulse duration while the electrodes made with Cu-W showed 10% wear for the same conditions. Palanikumar *et al.* [29] conducted tests to investigate the process parameters of machining WC/Co cemented carbide composite materials by EDM process. Various machining parameters were also analyzed and the effect of MRR and surface quality by these parameters was observed. In a similar study, Hanaoka *et al.* [30] investigated the effect of EDM process which used asiisting electode on the discharge behavior of composite made of Si_3N_4 CMs. In the conclusion, it was stated that the results of the MRR of the insulating materials were better than the results of conductive materials. Gadalla and Tsai [31] conducted some experimental studies to demonstrate the effect on the content of the cobalt and the size of the grains on various machining characteristics of the cemented tungsten carbide. It was observed that the rate of MRR can be varied by the percentage of cobalt. By increasing the cobalt percentage the MRR will also improve and also increase the EDM process stabilization. Lin *et al.* [32] conducted some investigations and tried to form a link between various parameters like MRR, surface roughness, cracks on surface with the rate of electrical discharge. In the results it was observed that by increasing the discharge of the electrical energy will significantly increase the MRR and surface roughness of the material formed by EDM process.

Song *et al.* [33] conducted an experiment to investigate the effect of EDM process on composite made of WC-Co with 10.5 5 of Co and using water spray as the dielectric medium. In the reults, it was observed that no corrosoion was reported by using water spray dielectric medium and the MRR was incraesed along with significant reduction in wear of the electrode. Yu *et al.* [34] investigated the experiments on cemented carbides using two types of EDM wiz dry EDM and conventional oil based EDM. In the results, it was observed that the dry EDM showed better results with high machining speeds and low wear of electrodes as compared with conventional oil based EDM. It was further stated that dry EDM can also be used for 3-Dimensional machining of the carbides used as workpiece material. In another study Kumar *et al.* [35] investigated the effects on some machining parameters like pulse on and off time, current of pulse and voltage by EDM process on composite made of Al635i with 5 wt. % of Sic and 5 wt. % of B_4C. Tool wear, surface roughness and power consumption were observed and it was stated that by increase in in the electric discharge and pulse current there are significant chances of increase in the wear of the tool. In a similar study Patel *et al.* [36] optimized the surface roughness parameters of composite materials made of Al_2O_3-SiC_w-TiC with EDM process. Improvement in mechanical properties was achieved by whiskers of SiC and particles of TiC into Al_2O_3. In the results it was concluded that surface methodology method was useful for obtaining better surface finish of the formed composite by the EDM process. Selvarajan *et al.* [37] reviewed various parameters of the compsoite material processed by EDM that are used in industrial applications. Various factors that are required to gain the better output were discussed and it was concluded that out of the studied materials $A_{12}O_3$, SiC and TiC are in demand for most of the applications due to their easiness in machining. Kangarajan [38] conducted a research and compared the results of wear mechanism of WC-Co composite materials made by powder metallurgy process and EDM process. Probabilistic neural networks were used to draw the wear mechanism maps of the different processes. The state of the wear was then divided into three parts: mild, severe and ultra-severe wear. These conditions were based on various factors like wear rate and surface roughness of the material. It was also observed that the mechanical properties of surfaces formed by EDM process were damaged by the presence of bubbles and craters arose through electric discharge. It was concluded from research that the state of ceramic wear was briefly recognized by mild wear, severe wear and ultra-severe wear from the viewpoints of wear surface roughness and wear rate. Electrical sparks generated craters, bubbles and cracks on the machined surface which deteriorated the surface finish and mechanical properties of the EDM WC-Co surface and should be removed for surface sensitive applications. Dinesh *et al.* [39] conducted a study and investigated the effect of different process parameters on various points such as feed rate of electrode; pulse on/ off time and electric discharge. In the results, it was observed the current was the main factors that significantly affect the MRR and surface roughness of the composite materials processed by EDM. Aidil *et al.* [40] conducted studies and critically examined various paremets that effect various maching factors effecting the EDM process. It was observed that the rate of MRR and the surface roughness of the material were decreased with increase in the pulse on time due to the expansion of plasma channel thus resulting in less removal of the material. Similarly, the values of the same parameters were increased by increase in pulse off time.

Table 2. Review on Electric Discharge machining of Ceramic composite material.

S.No	Author	Work Material	Input Parameters and Range	Optimized Result
1.	Janmanee and Muttamara [41].	WC-10%Co.	Discharge current, (1.5-75A). Off time (2-1600 µs). Open circuit voltage (90-250 V).	MRR: 2.7mm^3/min. EWR:37.2 mm^3/min. Micro crack density: 183.9 µm/mm$^{2.}$
2.	Shah *et al.* [42].	WC-12%Co.	Material thickness (25.4-76.2 mm). Open voltage (75-105V). Pulse on time (3-5µs). Pulse off time (20-30µs). Servo voltage (40-60V). Wire feed velocity (30-90mm/s). Wire tension (1000-2200g).	MRR: 13.64 mm^3/min. Kerf: 290.6 µm. SR: 1.11 µm.
3.	Singh *et al.* [43].	WC-10%Co.	Wheel speed (700-1100rpm). Current (4-8A). Pulse on time (50-150µs). Duty factor (0.57-0.70).	MRR: 0.3845 mm^3/min. WWR: 0.007042 g/min. ASR: 3.606 µm.
4.	Jangra *et al.* [44].	WC-6%Co.	Peak current (80-120A). Pulse on time (108-122µs). Pulse off time (30-50µs). Wire tension (6-10N). Dielectric flow rate (4-10LM-1).	MRR: 2.52 mm/min. SR: 0.88 µm.
5.	Vogeler *et al.* [27].	ZrO$_2$ -TiN.	Wire EDM, Operation: Melting, evaporation & Chemical decomposition.	Reduces for every cutting dimension, Bending strength does not changed.

(Table 2) cont.....

S.No	Author	Work Material	Input Parameters and Range	Optimized Result
6.	Vogeler *et al.* [45].	ZrO_2 -TiN.	Wire EDM, Operation: Melting, evaporation & Chemical decomposition.	Finish cutting seems not be related with flexural strength.
7.	Schubert *et al.* [46].	Si_3N_4 -TiN and Alumina Toughened Zirconia (ATZ).	Micro EDM, Open circuit voltage, discharge type.	The ablation behavior of Si3N4 -TiN enables 200% of MRR compared to ATZ.
8.	Yeakub *et al.* [47].	zirconia (titanium carbide.powder mixed with the kerosene).	Micro EDM, Gap voltage, capacitance.	The factor, which affected the most to MRR, was capacitance. 86 V and 1.0 nF are optimum for reaching maximum MRR.

Saxena *et al.* [48] investigated the effect of machining on SiC with EDM process. In the results obtained from the SEM images, it was clearly observed that the mechanism of the removal of the material was melting and evaporation and the migration of the material was also two directional.

Nihal *et al.* [49] used hydroxyapatite powder mixed to machine Ti6Al4v workpiece. Group found that random removal of workpiece and micro-cracks were formed when pulse time 100µs and pulse current 12A. Other non-contact processes such as laser machining, ultrasonic machining [50, 51], and abrasive water jet machining [52] also had their advantages and disadvantages as reported by different researchers.

Yoo *et al.* have successfully doped yttrium nitrate (YN) with SiC and showed that the EDM process can be used for machining these ceramics [53]. Yeakub *et al.* [54] investigated the micro-EDM of non-conductive zirconia with titanium carbide powder. In this study, two parameters were varied: capacitance and gap voltage.

Singh *et al.* [55] carried out an experimental investigation on the powder mixed EDM of cobalt-bonded tungsten carbide (WC-Co). In his experiment, graphite powder mixed dielectric was used. Few studies were focused on predictive modeling to understand the effect of EDM on the machined product [47]. Table **2** shows a consolidated review electric discharge machining of ceramic composite material.

FUTURE PROSPECT

On the basis of the various studies reviewed in this chapter, it has been observed that there is strong need to investigate the effect of various operational parameters on some nanostructured WC. Micro investigations of EDM, for CMs, are seen as major future scope in this field. There is need to investigate the formation of secondary layers which are conductive in nature with the help of some advanced FEM techniques.

CONCLUSION

Applications based on composites fabricated by CMs have become important part of our modern lifestyle. In this article, EDM process and its role in machining of composite made up of CMs and tungsten carbide has been reviewed. It is concluded from the above extensive review that EDM process is capable of machining CMs like TiN, SiC and TiB_2 which are highly electrically conductive in nature. It is also observed that various machining parameters like spark intensity and current effect the performance of the machining of the workpiece. As these parameters get increased, there is also a significant increase in the MRR and surface roughness. The result of pulse time is studied with respect to MRR. MRR and surface roughness decreases with increase in pulse time and thus increasing the instability of the machining of materials.

Along with this, effect on characteristics of EDM machined surfaces by different process parameters has also been observed and some studies have also been discussed stating the after machining effects on CMs [27, 45].

CONSENT FOR PUBLICATION

Not applicable.

CONFLICT OF INTEREST

The authors confirm that this chapter content has no conflict of interest.

ACKNOWLEDGEMENTS

The authors would like to express their sincere thanks to the editor and anonymous reviewers for their time and valuable suggestions.

REFERENCES

[1]   I. Puertas, C.J. Luis, and L. Álvarez, "Analysis of the influence of EDM parameters on surface quality, MRR and EW of WC-Co", *J. Mater. Process. Technol.,* vol. 153–154, no. 1–3, pp. 1026-1032, 2004. [http://dx.doi.org/10.1016/j.jmatprotec.2004.04.346]

[2]   A.S. Walia, V. Srivastava, V. Jain, and S.A. Bansal, "Effect of TiC Reinforcement in the Copper Tool on Roundness During EDM Process Lect. notes", *Mech. Eng.,* pp. 125-135, 2020.

[3]   G. Singh, and D. Parkash Dhiman, *Review: Parametric Optimization of Edm Machine Using Taghuchi & Anova Technique,* pp. 783-788, 2016.*Int. Res. J. Eng. Technol,* pp. 783-788, 2016.

[4]   B. Bhattacharyya, B.N. Doloi, and S.K. Sorkhel, "Experimental investigations into electrochemical discharge machining (ECDM) of non-conductive ceramic materials", *J. Mater. Process. Technol.,* vol.

95, no. 1–3, pp. 145-154, 1999.
[http://dx.doi.org/10.1016/S0924-0136(99)00318-0]

[5]   Y.H. Liu, X.P. Li, R.J. Ji, L.L. Yu, H.F. Zhang, and Q.Y. Li, "Effect of technological parameter on the process performance for electric discharge milling of insulating Al2O3 ceramic", *J. Mater. Process. Technol.,* vol. 208, no. 1–3, pp. 245-250, 2008.
[http://dx.doi.org/10.1016/j.jmatprotec.2007.12.143]

[6]   M. Kiyak, and O. Çakir, "Examination of machining parameters on surface roughness in EDM of tool steel", *J. Mater. Process. Technol.,* vol. 191, no. 1–3, pp. 141-144, 2007.
[http://dx.doi.org/10.1016/j.jmatprotec.2007.03.008]

[7]   V. Yadav, V.K. Jain, and P.M. Dixit, "Thermal stresses due to electrical discharge machining", *Int. J. Mach. Tools Manuf.,* vol. 42, no. 8, pp. 877-888, 2002.
[http://dx.doi.org/10.1016/S0890-6955(02)00029-9]

[8]   A. Schubert, H. Zeidler, M.H. Oschätzchen, J. Schneider, and M. Hahn, ""Micro-EDM using Ultrasonic Vibration and Approaches for Machining of Nonconducting Ceramics," Strojniški Vestn. –", *Jixie Gongcheng Xuebao,* vol. 59, no. 3, pp. 156-164, 2013.
[http://dx.doi.org/10.5545/sv-jme.2012.442]

[9]   A. Schoth, R. Förster, and W. Menz, "Micro wire EDM for high aspect ratio 3D microstructuring of ceramics and metals", *Microsyst. Technol.,* vol. 11, no. 4–5, pp. 250-253, 2005.
[http://dx.doi.org/10.1007/s00542-004-0399-y]

[10]   C.J. Luis, I. Puertas, and G. Villa, "Material removal rate and electrode wear study on the EDM of silicon carbide", *J. Mater. Process. Technol.,* vol. 164–165, pp. 889-896, 2005.
[http://dx.doi.org/10.1016/j.jmatprotec.2005.02.045]

[11]   W. König, D.F. Dauw, G. Levy, and U. Panten, ""EDM-Future Steps towards the Machining of Ceramics," CIRP Ann. -", *Manuf. Technol.,* vol. 37, no. 2, pp. 623-631, 1988.
[http://dx.doi.org/10.1016/S0007-8506(07)60759-8]

[12]   A. Schubert, H. Zeidler, N. Wolf, and M. Hackert, "Micro electro discharge machining of electrically nonconductive ceramics", *AIP Conf. Proc.,* vol. 1353, pp. 1303-1308, 2011.
[http://dx.doi.org/10.1063/1.3589696]

[13]   B. Lauwers, J.P. Kruth, W. Liu, W. Eeraerts, B. Schacht, and P. Bleys, "Investigation of material removal mechanisms in EDM of composite ceramic materials", *J. Mater. Process. Technol.,* vol. 149, no. 1–3, pp. 347-352, 2004.
[http://dx.doi.org/10.1016/j.jmatprotec.2004.02.013]

[14]   C. Martin, B. Cales, P. Vivier, and P. Mathieu, "Electrical discharge machinable ceramic composites", *Mater. Sci. Eng. A,* vol. 109, no. C, pp. 351-356, 1989.
[http://dx.doi.org/10.1016/0921-5093(89)90614-X]

[15]   A. Schubert, H. Zeidler, R. Kühn, and M. Hackert-Oschätzchen, "Microelectrical discharge machining: A suitable process for machining ceramics", *J. Ceram.,* vol. 2015, pp. 1-9, 2015.
[http://dx.doi.org/10.1155/2015/470801]

[16]   S.H. Lee, and X.P. Li, "Study of the effect of machining parameters on the machining characteristics in electrical discharge machining of tungsten carbide", *J. Mater. Process. Technol.,* vol. 115, no. 3, pp. 344-358, 2001.
[http://dx.doi.org/10.1016/S0924-0136(01)00992-X]

[17]   D. Kanagarajan, R. Karthikeyan, K. Palanikumar, and J.P. Davim, "Application of goal programming technique for electro discharge machining (EDM) characteristics of cemented carbide (WC/Co)", *Int. J. Mater. Prod. Technol.,* vol. 35, no. 1–2, pp. 216-227, 2009.
[http://dx.doi.org/10.1504/IJMPT.2009.025228]

[18]   K. Jangra, S. Grover, F.T.S. Chan, and A. Aggarwal, "Digraph and matrix method to evaluate the machinability of tungsten carbide composite with wire EDM", *Int. J. Adv. Manuf. Technol.,* vol. 56,

no. 9–12, pp. 959-974, 2011.
[http://dx.doi.org/10.1007/s00170-011-3234-5]

[19]   G.N. Levy, and R. Wertheim, ""EDM-Machining of Sintered Carbide Compacting Dies," CIRP Ann. - ", *Manuf. Technol.,* vol. 37, no. 1, pp. 175-178, 1988.
[http://dx.doi.org/10.1016/S0007-8506(07)61612-6]

[20]   K. Bonny, P. De Baets, J. Vleugels, O. Van der Biest, B. Lauwers, and W. Liu, "EDM machinability and dry sliding friction of WC-Co cemented carbides", *Int. J. Manuf. Res.,* vol. 4, no. 4, pp. 375-394, 2009.
[http://dx.doi.org/10.1504/IJMR.2009.028536]

[21]   A.M. Gadalla, and N.F. Petrofes, *Materials and Manufacturing Processes SURFACES OF ADVANCED CERAMIC COMPOSITES FORMED BY ELECTRICAL DISCHARGE MACHINING,* 2007.

[22]   M.N. Farooqui, and N.G. Patil, "A perspective on shaping of advanced ceramics by electro discharge machining", *Procedia Manuf.,* vol. 20, pp. 65-72, 2018.
[http://dx.doi.org/10.1016/j.promfg.2018.02.009]

[23]   J.R. Gamage, A.K.M. DeSilva, C.S. Harrison, and D.K. Harrison, "Process level environmental performance of electrodischarge machining of aluminium (3003) and steel (AISI P20)", *J. Clean. Prod.,* vol. 137, no. 3003, pp. 291-299, 2016.
[http://dx.doi.org/10.1016/j.jclepro.2016.07.090]

[24]   B. Lauwers, J. Vleugels, O. Malek, K. Brans, and K. Liu, "Electrical discharge machining of composites", *Mach. Technol. Compos. Mater. Princ,* pp. 202-241, 2011.
[http://dx.doi.org/10.1016/B978-0-85709-030-0.50008-4]

[25]   F. Bucciotti, M. Mazzocchi, and A. Bellosi, "Perspectives of the Si3N4-TiN ceramic composite as a biomaterial and manufacturing of complex-shaped implantable devices by electrical discharge machining (EDM)", *J. Appl. Biomater. Biomech.,* vol. 8, no. 1, pp. 28-32, 2010.
[PMID: 20740419]

[26]   N. Mohri, Y. Fukuzawa, T. Tani, N. Saito, and K. Furutani, ""Assisting Electrode Method for Machining Insulating Ceramics," CIRP Ann. -", *Manuf. Technol.,* vol. 45, no. 1, pp. 201-204, 1996.
[http://dx.doi.org/10.1016/S0007-8506(07)63047-9]

[27]   F. Vogeler, B. Lauwers, and E. Ferraris, *Analysis of Wire-EDM Finishing Cuts on Large Scale ZrO2-TiN Hybrid Spark Plasma Sintered Blanks,* 2016.
[http://dx.doi.org/10.1016/j.procir.2016.02.284]

[28]   S. Sano, K. Suzuki, W. Pan, M. Iwai, Y. Murakami, and T. Uematsu, "Forming fine V-grooves on a tungsten carbide workpiece with a PCD electrode by EDM", *Key Eng. Mater.,* vol. 329, pp. 631-636, 2007.
[http://dx.doi.org/10.4028/www.scientific.net/KEM.329.631]

[29]   K. Palanikumar, *Machining and machine-tools.* Woodhead Publishing Limited, 2013.

[30]   D. Hanaoka, Y. Fukuzawa, C. Ramirez, P. Miranzo, M.I. Osendi, and M. Belmonte, "Electrical discharge machining of ceramic/carbon nanostructure composites", *Procedia CIRP,* vol. 6, pp. 95-100, 2013.
[http://dx.doi.org/10.1016/j.procir.2013.03.033]

[31]   A.M. Gadalla, and W. Tsai, "Electrical Discharge Machining of Tungsten Carbide☐Cobalt Composites", *J. Am. Ceram. Soc.,* vol. 72, no. 8, pp. 1396-1401, 1989.
[http://dx.doi.org/10.1111/j.1151-2916.1989.tb07660.x]

[32]   Y.C. Lin, L.R. Hwang, C.H. Cheng, and P.L. Su, "Effects of electrical discharge energy on machining performance and bending strength of cemented tungsten carbides", *J. Mater. Process. Technol.,* vol. 206, no. 1–3, pp. 491-499, 2008.
[http://dx.doi.org/10.1016/j.jmatprotec.2007.12.056]

[33]　K.Y. Song, D.K. Chung, M.S. Park, and C.N. Chu, "Water spray electrical discharge drilling of WC-Co to prevent electrolytic corrosion", *Int. J. Precis. Eng. Manuf.,* vol. 13, no. 7, pp. 1117-1123, 2012.
[http://dx.doi.org/10.1007/s12541-012-0147-7]

[34]　Z.B. Yu, T. Jun, and K. Masanori, "Dry electrical discharge machining of cemented carbide", *J. Mater. Process. Technol.,* vol. 149, no. 1–3, pp. 353-357, 2004.
[http://dx.doi.org/10.1016/j.jmatprotec.2003.10.044]

[35]　S.S. Kumar, M. Uthayakumar, S.T. Kumaran, and P. Parameswaran, "Electrical discharge machining of Al(6351)-SiC-B4C hybrid composite", *Mater. Manuf. Process.,* vol. 29, no. 11–12, pp. 1395-1400, 2014.
[http://dx.doi.org/10.1080/10426914.2014.952024]

[36]　K.M. Patel, P.M. Pandey, and P. Venkateswara Rao, "Determination of an optimum parametric combination using a surface roughness prediction model for EDM of Al 2 O 3 /SiC w /TiC ceramic composite", *Mater. Manuf. Process.,* vol. 24, no. 6, pp. 675-682, 2009.
[http://dx.doi.org/10.1080/10426910902769319]

[37]　L. Selvarajan, J. Rajavel, V. Prabakaran, B. Sivakumar, and G. Jeeva, "A Review Paper on EDM Parameter of Composite material and Industrial Demand Material Machining", *Mater. Today Proc.,* vol. 5, no. 2, pp. 5506-5513, 2018.
[http://dx.doi.org/10.1016/j.matpr.2017.12.140]

[38]　D. Kanagarajan, *Influence of Electro Discharge Machining, Micro Structural and Mechanical Properties on Wear Behaviour of WC-30% Co Composites,* 2016.

[39]　S. Dinesh Kumar, M. Ravichandran, S. V. Alagarsamy, M. Meignanamoorthy, and S. Sakthivelu, "Effect of EDM process parameters on material removal rate and surface roughness of metal matrix composites: A review", *Mater. Today Proc,* vol. xxxx, 2019.
[http://dx.doi.org/10.1016/j.matpr.2019.06.725]

[40]　A. R. M. Aidil, M. Minhat, and N. I. S. Hussein, "Current research trends in Wire Electrical Discharge Machining (WEDM): A review,", *J. Adv. Manuf. Technol,* vol. 12, pp. 11-24, 2018.

[41]　P. Janmanee, and A. Muttamara, "Optimization of electrical discharge machining of composite 90WC-10Co base on taguchi approach", *Eur. J. Sci. Res.,* vol. 64, no. 3, pp. 426-436, 2011.

[42]　A. Shah, N.A. Mufti, D. Rakwal, and E. Bamberg, "Material removal rate, kerf, and surface roughness of tungsten carbide machined with wire electrical discharge machining", *J. Mater. Eng. Perform.,* vol. 20, no. 1, pp. 71-76, 2011.
[http://dx.doi.org/10.1007/s11665-010-9644-y]

[43]　G.K. Singh, V. Yadava, and R. Kumar, "Diamond face grinding of WC-Co composite with spark assistance: Experimental study and parameter optimization", *Int. J. Precis. Eng. Manuf.,* vol. 11, no. 4, pp. 509-518, 2010.
[http://dx.doi.org/10.1007/s12541-010-0059-3]

[44]　K. Jangra, S. Grover, and A. Aggarwal, "Simultaneous optimization of material removal rate and surface roughness for WEDM of WCCo composite using grey relational analysis along with Taguchi method", *Int. J. Ind. Eng. Comput.,* vol. 2, no. 3, pp. 479-490, 2011.
[http://dx.doi.org/10.5267/j.ijiec.2011.04.005]

[45]　F. Vogeler, A. Migdal, B. Lauwers, and E. Ferraris, "The Effect of Wire -EDM Processing on the Flexural Strength of Large Scale ZrO2-TiN", *Procedia CIRP,* vol. 45, pp. 179-182, 2016.
[http://dx.doi.org/10.1016/j.procir.2016.02.333]

[46]　A. Schubert, H. Zeidler, R. Kühn, M. Hackert-Oschätzchen, S. Flemmig, and N. Treffkorn, *Investigation of Ablation Behaviour in Micro-EDM of Nonconductive Ceramic Composites ATZ and Si3N4-TiN,* 2016.
[http://dx.doi.org/10.1016/j.procir.2016.02.309]

[47]　"A. goleman, daniel; boyatzis, Richard; Mckee, No Title No Title", *J. Chem. Inf. Model.,* vol. 53, no.

9, pp. 1689-1699, 2019.
[http://dx.doi.org/10.1017/CBO9781107415324.004]

[48]   K. K. Saxena, S. Agarwal, and S. K. Khare, *Surface Characterization, Material Removal Mechanism and Material Migration Study of Micro EDM Process on Conductive SiC,* 2016.
[http://dx.doi.org/10.1016/j.procir.2016.02.267]

[49]   N. Ekmekci, and B. Ekmekci, "Electrical Discharge Machining of Ti6Al4V in Hydroxyapatite Powder Mixed Dielectric Liquid", *Mater. Manuf. Process.,* vol. 31, no. 13, pp. 1663-1670, 2016.
[http://dx.doi.org/10.1080/10426914.2015.1090591]

[50]   R.P. Singh, and S. Singhal, "An experimental study on rotary ultrasonic machining of Macor ceramic", *Proc. Inst. Mech. Eng., B J. Eng. Manuf.,* vol. 232, no. 7, pp. 1221-1234, 2018.
[http://dx.doi.org/10.1177/0954405416666897]

[51]   R.P. Singh, and S. Singhal, ""Experimental investigation of machining characteristics in rotary ultrasonic machining of quartz ceramic," Proc. Inst. Mech. Eng. Part L J", *Mater. Des. Appl.,* vol. 232, no. 10, pp. 870-889, 2018.
[http://dx.doi.org/10.1177/1464420716653422]

[52]   A.M. Deris, A.M. Zain, and R. Sallehuddin, "A note of hybrid GR-SVM for prediction of surface roughness in abrasive water jet machining: a response", *Meccanica,* vol. 52, no. 8, pp. 1993-1994, 2017.
[http://dx.doi.org/10.1007/s11012-016-0551-7]

[53]   H.K. Yoo, J.H. Ko, K.Y. Lim, W.T. Kwon, and Y.W. Kim, "Micro-electrical discharge machining characteristics of newly developed conductive SiC ceramic", *Ceram. Int.,* vol. 41, no. 3, pp. 3490-3496, 2015.
[http://dx.doi.org/10.1016/j.ceramint.2014.10.175]

[54]   M. A. Moudood, A. Sabur, M. Y. Ali, and I. H. Jaafar, ">Effect of Gap Voltage and Pulse-On Time on Material Removal Rate for Electrical Discharge Machining of Al_2O_3", *Adv. Mater. Res,* pp. 3-6, 2015.
[http://dx.doi.org/10.1016/j.ceramint.2014.10.175]

[55]   J. Singh, and R.K. Sharma, "Assessing the effects of different dielectrics on environmentally conscious powder-mixed EDM of difficult-to-machine material (WC-Co)", *Front. Mech. Eng.,* vol. 11, no. 4, pp. 374-387, 2016.
[http://dx.doi.org/10.1007/s11465-016-0388-8]

CHAPTER 7

Current Developments in Machining of Titanium Based Alloys using Wire EDM

Kamaljit Singh[1] and **Virat Khanna[2,*]**

[1] *Chandigarh Engineering College, Mohali, India*

[2] *Department of Mechanical Engineering, Maharaja Agrasen University, Baddi, Himachal Pradesh, India*

Abstract: Demand for advanced materials, possessing, and better properties is always there in the multi-disciplinary fields of automotive, aerospace and medical applications requires advanced machining processes. Wire electric discharge machining (WEDM) is a promising non-traditional machining method utilized extensively to machine materials with varying hardness and intricate shapes with great accuracy and precision. A WEDM is a non-contact process, involves the removal of material in the form of spark erosions. DC electrical sparks are constantly generated between the wire electrode and the workpiece to achieve desired shape on the component. The input variable such as pulse-on-time, pulse-off-time, wire speed, servo voltage and flushing pressure can be tuned and controlled in order to optimize the machining process responses. The aim of this study is to identify the significant effects of process parameters on MRR, TWR, SR and surface morphology of titanium nickel memory alloys. Additionally, areas have been recognized that require intense research considerations. Future aspects of WEDM are also presented.

Keywords: Artificial neural network (ANN), Design of experiment (DOE), Electron back scatter diffraction (EBSD), Energy dispersive spectroscopy (EDS), Metal removal rate (MRR), Scanning electron microscopy (SEM), Surface roughness (SR), Transmission electron microscopy (TEM), Wire electric discharge machining (WEDM).

INTRODUCTION

Manufacturing industry, nowadays, follows a trend towards high-precision machining. This can easily be found among the demanding markets of micro-mechanical components, electronics, and aerospace industries [1]. As newer and more exotic materials have been developed in the past few decades, conventional machining operations have reached their limits.

* **Corresponding author Virat Khanna**: Department of Mechanical Engineering, Maharaja Agrasen University, Baddi, Himachal Pradesh, India; E-mail: khanna.virat@gmail.com

As industry advances, relatively more complicated shaped jobs are required to be manufactured [2]. WEDM was introduced in late nineties and gained great popularity since then in manufacturing industry. This is primarily due to its great ability in machining hard to cut, high strength to weight ratio materials with the capacity of producing complex shapes and profiles. Furthermore, WEDM can produce a fine, dimensionally accurate surface with high precision. Wire electrical discharge machining is a spark erosion process used to produce complex two- and three-dimensional shapes through electrically conductive workpieces. WEDM differs from conventional electrical discharge machining. A thin wire performs as the electrode and finishes the cutting operation [3].

Wire EDM is a special form of the traditional EDM process in which the electrode is a continuously moving. Material is eroded from the workpiece by a series of discrete sparks between the workpiece and the wire electrode (tool) separated by a thin film of dielectric fluid (deionised water). Wire EDM provides the best alternative or sometimes the only alternative for machining conductive, exotic, high strength and temperature resistive (HSTR) materials with the scope of generating complex geometry. It has proved to have tremendous potential in its applicability in the present-day metal cutting industry for achieving a considerable dimensional accuracy, surface finish and contour generation features of products or parts [4].

Machining of titanium alloy in the minimum time and with maximum precision is an important issue in all application especially biomedical [5]. Titanium and its alloys are classified as difficult to machine materials as they possess high toughness and strength even at extreme temperatures. Due to properties like light weight, extraordinary resistance to corrosion *etc.*, Ti alloys *etc.* find diversified and wide spectrum of applications in biomedical, automobile, aerospace, chemical field, electronic, gas and food industry. Moreover, titanium alloys incredibly durable and long lasting. When titanium cages, rods, plates and pins are inserted into the body, they can remain unaffected for more than twenty years [6]. Machining of titanium alloy, with conventional methods, is very inconvenient because of the problems such as immense tool wear; high temperature requirement during cutting *etc.* Therefore, these alloys can be machined with WEDM as it is more suitable for such materials. The non-ferromagnetic property of Ti alloys is an additional advantage that allows patients with different types of titanium inserts within their bodies, to be safely analysed under MRIs and NMRIs [7]. Some implants made up of titanium are shown in Fig. (**1**).

<table>
<tr><td>a. Orthopedic implants</td><td>b. Dentist implants</td></tr>
</table>

Fig. (1). Medical implants made up of Titanium alloys [7].

WIRE ELECTRIC DISCHARGE MACHINING (WEDM)

While the material removal mechanisms of EDM and WEDM are similar, their functional characteristics are not identical [8]. In WEDM process electrical sparks are produced between a thin, travelling wire which acts as the tool electrode and the workpiece to erode the material in order to generate desired shape. The Schematic diagram of WEDM is shown in Fig. **(2)**.

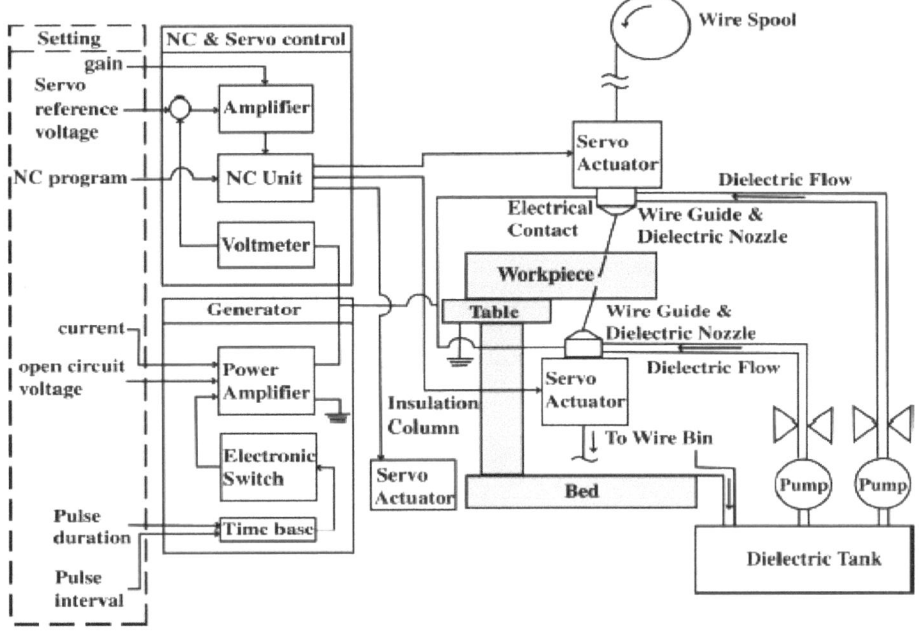

Fig. (2). Schematic diagram of a WEDM system [9].

During operation, the wire is continuously fed from a wire spool and is guided by lower and upper nozzle in wire electric discharge machine while the workpiece remains stationary. The materials usually used are conductive in nature such as copper, Brass, Zinc coated brass, Diffused brass, Molybdenum *etc.* The electric potential isapplied between the wire electrode and workpiece, where electrode is acting as a cathode and workpiece is act as anode while maintaining a small gap between them.

The dielectric fluid, such as de-ionized water, is also present between the wire electrode and workpiece. The electron moves from the cathode and its strike on to the molecule of de-ionized water and releases the number of electrons. These electrons again strike on to the next molecule of the de-ionized water. The process goes on continuously and simultaneously positive ions move towards the cathode. In this way, a large number of electrons are moved towards the anode (workpiece). The kinetic energy of electrons is converted into thermal energy and spark is generated. This spark energy increases the temperature of the workpiece to about 10,000°C. Due to this, melting and evaporation of workpiece material are taken place [10]. Filtered dielectric fluid is pumped and streamed continuously to enclose both the workpiece and the wire electrode. The electrical energy pulses passing through the electrode wire, at the machining zone, vaporize the dielectric fluid to blow off tiny portions of the workpiece and the wire. As, wire is continuously fed by the wire spool, every time a new section of wire acts as the electrode when old section of the wire is cut off and becomes scrap. However, the wire-workpiece gap is maintained and controlled to produce an accurate cut surface. Moreover, the dielectric fluid acts as an insulator so that high energy in the gap can be produced for the generation of spark. Also, it collects and carries away the debris and reduces heat that is produced during erosion.

Wire-cut EDM is used when low residual stresses are desirable as it does not need high cutting forces for the removal of the material. If the energy per pulse is relatively low, little change in the mechanical properties of a material is expected due to these low residual stresses. Although, in conventional machining processes, material that has not been stress- relieved can distort in the during service life. Due to the inherent properties of the process, wire EDM can easily machine complex parts and precision components.

WEDM Process Parameters

various electrical and non-electrical process parameters, their inter-machining relationship and performance measures of WEDM are shown in Fig. (**3**). It is important to select adequate levels for these parameters to achieve economic machining conditions. The input variables can be set, considering the effects of these on the output variables, to attain optimum machining environment.

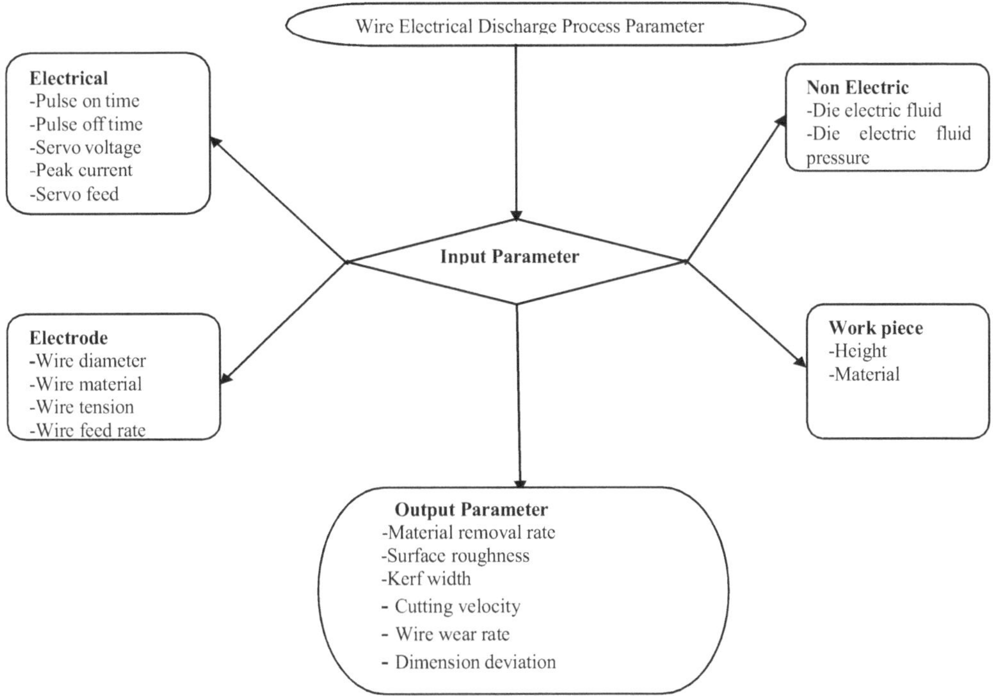

Fig. (3). WEDM process and respond parameter [11].

WIRE Electrode Properties

The machining performances of the WEDM, in general, rely on combination of various physical, chemical, electrical and mechanical properties of the tool electrode. Thus, wire electrode should have excellent properties for achieving high-speed cutting with great accuracy and precision. In Table **1** the list distinct properties of WEDM wires.

Table 1. Essential properties of WEDM wire [12].

S.No	Property	Description
1.	Conductivity	Low wire conductivity results in a voltage drop and associated energy loss over the distance from the power feed to the cutting point
2.	Tensile Strength	Hard wires at tensile strength of 900MPa are used for most of work, while half hard around 490MPa and soft wires below 440Mpa are used for taper cuts for angle more than 5°.

(Table 1) cont.....

S.No	Property	Description
3.	Elongation	Hard wires have considerably less elongation than half hard wires. A brittle wire might break at first overload condition, while a more ductile wire is expected to accept a temporary overload
4.	Melting Point	Wire electrode should be resistant to the rapid melting by the sparks
5.	Geometrical	Wires ranging in diameters from 0.02 to0.36 mm are generally available for WEDM. An increase in diameter of wire results in an increase of pulse energy supplied to the working gap, Thus increasing the overall material removal rate

Fig. (**4**) illustrates the increase in wire diameter with the development in the wire electrode for improvement in the WEDM process [13].

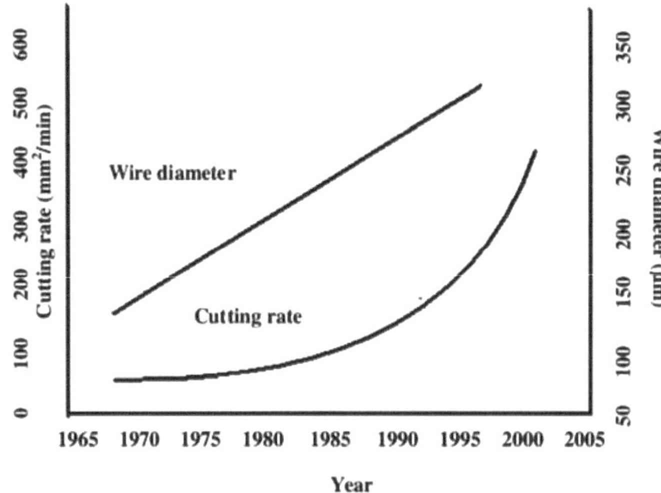

Fig. (4). Higher cutting rate machines using thick wires [13].

Advancements in WEDM Process

WEDM has proven itself as the advanced technique for machining hard materials with complex geometry configuration with high precision rate. However, other machining processes can be integrated together with WEDM for better machining. Table **2** shows the modern WEDM techniques developed in recent years.

Table 2. Hybrid WEDM Techniques.

S.No.	Hybrid WEDM
1.	Wire EDM Grinding
2.	Micro WEDG
3.	Wire EDM Milling

(Table 2) cont.....

S.No.	Hybrid WEDM
4.	Wire EDM Turning
5.	Abrasive Wire EDM

Wire Electric Discharge Grinding

In order to produce cylindrical micro-parts, machining process like wire electric discharge grinding (WEDG) is preferred [14]. In WEDG process, micro rotary parts such as rods and pins are machined with high precision [15]. However, the WEDG process can fabricate micro rotational parts with curved contour profile precisely ; the machining efficiency is very limited. Besides, this technology depends on an expensive and dedicated micro EDM machine, which restricts the popularization of the technology in common industries [16].

Micro Wire Elcetro Discharge Grinding

In micro-WEDG the uses of the grinding wheel are replaced with micro wire which acts as an electrode. When fine rod needs to be ground, it acts as a workpiece and allowed to be rotated while the electrode (micro wire) is held stationary. This process utilizes the electrical sparks or thermal energy produced to remove the unwanted materials from the workpiece in order to make the desired shape [17]. However, wire breakage control is most critical aspect of micro-WEDM process from technical and economic grounds.

Wire Electric Discharge Turning

Wire electrical discharge turning (WEDT) process is one of the emerging non-traditional machining processes for the manufacture of micro- and axi-symmetric components [18]. In WEDT and WEDG, the erosion mechanism is similar to that of the WEDM process with a difference that in WEDG the workpiece is fed in longitudinal direction rather than in horizontally as in WEDT. Moreover, the characteristics of process variants of the WEDG and WEDT process are in the relative motion between the tool and workpiece electrodes and the feed.

Abrasive WEDM (AWEDM)

In this process, electrical sparks along-with abrasives abrade the workpiece surface. Higher material removal rate and better surface finish can be accomplished by integrating WEDM with abrasives of fine size.

DEVELOPMENTS IN WEDM OF TITANIUM BASED ALLOYS

Srinivasan and Palani [19] examined the silicon nitride-titanium nitride (Si_3N_4-TiN) ceramic composite to characterize its surface integrity and fatigue performance in WEDM. Different types of defects, such as micro-cracks, micro-poles, globules, droplets and surface craters, were found during SEM on the WEDM surface of the ceramic composite. Moreover, the fatigue life of Wire EDM cut surface decreased by 25-35% than that of polished surface. Also, the hardness of the polished surface was much better than of the machined surface.

Takale and Chougle [20] mentioned that Ti49.4Ni50.6 shape memory alloy have low young's modulus which makes it excellent for novel orthopedic implants. However, Ti-Ni yielded poor surface quality and low dimensional accuracy when it was machined with conventional methods. Therefore, WEDM can be an alternative to achieve high dimensional accuracy for machining Ti-Ni memory alloy. MRR and SR of Ti-Ni alloy were investigated for varying input parameter of WEDM. The input parameter selected were pulse on time, pulse off time, spark gap set voltage, wire feed and wire tension. GRA was employed for multi objective optimization. To study metallurgical changes on the machined surfaces, EDS analysis was performed.

Singh and Misra [21] investigated nimonic 263 in WEDM under varying input variables. The design of the experiment was based on Box-Behnken design technique while parametric optimization for surface roughness was performed using response surface methodology. Results from BPNN showed that BPNN was a valuable tool for the prediction of responses. It was identified in FE-SEM that WEDMed surface contained micro-cracks, craters, spherical droplets and lumps of debris.

Pramanik and Basak [22] presented the effect of input parameters on the fatigue life, crack propagation, surface roughness and machining during wire EDM of Ti-6Al-4V alloy. Machining parameters, pulse on time, pulse off time and wire tension were considered and influence of these on surface integritry were observed using SEM. The chief contributors observed for fatigue performance were nature and extent of recast layer and HAZ.

Kavimani *et al.* [23] used Taguchi coupled grey relation analysis to investigate the WEDM cutting parameters on metal matrix composite of magnesium reinforced with reduced grapheme oxide. L27 orthogonal array was applied to design the experiments and results were confirmed with the application of ANOVA. It was determined that reinforcement weight percentage and pulse on time were the most influential parameters among other input parameters such as pulse off time and wire feed rate for MRR and SR.

Chu *et al.* [24] discussed about a new feasible low frequency vibration assisted technique for generating microgroove on metal plates for high speed WEDM. The ridge width was significantly affected by the average current, vibration frequency and feed rate. However, low frequency vibration assisted WEDM can generate a smaller microgroove surface structure and higher processing efficiency as compared to traditional WEDM process.

Rajkumar *et al.* [25] addressed machining of aluminium metal matrix composites reinforced with nano alumina particle in Wire electric discharge machining. Voltage and pulse on time were found responsible input parameters for achieving higher MRR. Also, controlled pulse durations improved the surface integrity by reducing the machining defects.

Zhang *et al.* [26] analyzed and reduced the power consumption during machining of thin wall components in WEDM and studied the thermal deformation of the components. They utilized new micro-crack wire electrode to ameliorate machining characters. Also, the investigation results showed that micro crack wire can reduce the energy consumption and debris pollution at great level.

Sharma *et al.* [27] presented the optimization of input variables of WEDM while machining $Ni_{55.8}Ti$ shape memory alloy rich in Ni, specifically used in Bio-medical. L_{16} orthogonal array was selected using Taguchi DOE for conducting the study. Outcomes from the study confirmed that machining at optimum conditions results in better surface integrity and higher MRR.

Smirnov *et al.* [28] performed WEDM of 3y-TZP/Ta ceramic- metal composites and found that SPS Sintered composites have suitable electrical conductivity for WEDM process. Moreover, conclusion was made that high accuracy can be achieved while machining without much compromising with mechanical strength of the component.

Zhou *et al.* [29] fabricated and characterized the hydrophobic Ti_3SiC_2 surfaces in Wire electric discharge machining with regular and controllable micro-grooved structures. Relationships between contact angles, depth-width ratio and surface quality were also investigated and it was found that depth- width ration can be useful in predicting the parallel contact angle with a small margin of error.

Pramanik and Basak [30] mentioned the importance of uninterrupted machining in order to control energy and time consumption in WEDM for hard materials. Wire rupture for Ti-6Al-4V alloy workpiece was investigated to make the machining process highly sustainable. No workpiece material was detected on the tips of the wire, though, traces of oxides were formed because of high temperature oxidation were noticed on wire broken tips. Furthermore, at low temperature, it was observed, wire with high tension can rapture rapidly.

Thankachan *et al.* [31] predicted the performance measures in WEDM during machining of aluminium based alloys. Taguchi technique coupled with GRA was utilized to perform the investigation. Optimum MRR and Surface Roughness values were predicted based on the input control factors by developing artificial neural network approach. Outcomes were found efficient when compared, different models.

Nain *et al.* [32] used particle swarm optimization technique so as to improve the accuracy and productivity of wire electric discharge machining of super alloy Udimet-L605. Among other input variable, pulse on time was found to impact the cutting speed, wire wear ration and geometrical deviation extremely. The compositional modifications because of erosion and HAZ were studied with SEM and EDS.

Chaudhary *et al.* [33] studied the influence of varying wire tension on performance measures while machining AISI 304 in WEDM. It was noticed that wire tension had no influence on kerf width, but, affects the MRR, SR and micro-hardness significantly. While SR decreases with an increase in wire tension, slight improvement was recorded in case of MRR.

Gaitonde *et al.* [34] discussed the optimization of input variable of WEDM of HCHr steel employing integrating approach of RSM and differential evolution. L27 orthogonal array was selected based on Taguchi experiment design technique to carry out the investigation. More pulse on time resulted in higher MRR.

Goyal [35] represented the study of optimization of WEDM key input parameters for improved MRR and SR of Inconel 625 by employing normal and cryogenic zinc coated wires. MRR and SR were found considerably affected by pulse on time, tool electrode and current intensity. Microstructure of the machined surface was analyzed through SEM.

Mouralova *et al.* [36] distinguished the morphology and topography of WEDM structural material surfaces. SEM and contact less 3D profilometer were utilized to carry out the analysis. These instruments permitted the evaluation of profile and creating 3D surface image for further study.

Arikatla *et al.* [37] discussed the parametric optimization of WEDM process using RSM while machining Ti-6Al-4V alloy. The experiment was carried out with five input process parameters for improving MRR, kerf width and SR. surface properties were examined using SEM. Wire tension and servo voltage had high influence on surface roughness and enhanced the surface quality of machined surface.

Siddiqui and Ramkumar [38] suggested that magnesium which is an important material in biodegradable orthopedic implants applications can be machined in micro size complex shapes in micro-WEDM. AZ31 Mg alloy was selected to execute the experiments to set the relationship between various input parameters of micro WEDM. For optimum conditions, it was desirable to have medium voltage and high capacitance for high MRR.

Liu *et al.* [39] described the machining of Nithinol shape memory alloy in WEDM. Experiment was conducted to identify the effect of process parameters on the crystallography, composition and properties of white layer. TEM, X ray Diffraction and EBSD were used to characterize the properties of white layer on machined surface. In comparison to bulk material, the White layer showed higher nano-hardness due to oxide hardening.

Manjaiah *et al.* [40] studied the effect of wire electrode material during machining of titanium shape memory alloy in WEDM. Brass and Zinc Coated wires were used for the study. It was concluded that pulse on time, pulse off time were most impactful parameters along-with servo voltage in optimizing conditions for enhancing MRR and SR of machined material.

Sun *et al.* [41] discussed the characteristics of micro milling helical and corrugated micro end mill in low speed wire electric discharge machining. The design and mathematical simulation of wire electrode trajectory were carried out and effects of machining parameters on helical end mills were elucidated.

Table 3. Review on WEDM.

S.No.	Author	Optimization Technique	Work Material	Input Parameter	Outcomes
1.	Shakeri *et al.* [42]	ANN	Cementation alloy steel	Pulse Current, A (8-16). Frequency, kHz (40-60). Wire Speed, m/min (8-12). Servo speed, mm/min (4-8)	Increase in wire speed, pulse current and frequency leads to high MRR
2.	Goyal *et al.* [43]	ANN, RSM, SEM	$Ni_{49}Ti_{51}$	Current, A (4-8). Pulse on Time, µs (60-120). Pulse off time, µs (15-45). Wire tension, cm^2/gm (11-15). Wire feed rate, m/min (4-8)	Increased Pulse on time and Current increases MRR.

(Table 3) cont.....

S.No.	Author	Optimization Technique	Work Material	Input Parameter	Outcomes
3.	Suthan and Muralidharan [44]	Taguchi	316L	Pulse on Time, μs (30-32). Pulse off time, μs (5-7). Gap, mm/min (50-700	Higher Pulse on time results in Higher MRR
4.	Liao *et al.* [45]	Taguchi quality design, ANOVA and F-test	Alloy	Pulse generating circuit, (ac, dc). Conductivity of dielectric, (15-45). Capacitance, C (0-20). App. Voltage, V (100-150). Feed rate of the table, mm/min (2-4). Pulse off time, microsec (4-8). Resistance in circuit, ohm (25-75).	Pulse generating circuit, Voltage and resistance in circuit had significant effect on surface roughness
5.	Padmavathi *et al.* [46]	Taguchi Robust design approach	AISI 316 Stainless steel	Pulse on Time, μs (1-10). Pulse off time, μs (1-10). Peak Current, A (0-30) Wire tension, cm²/gm (200-1500). Wire feed rate, m/min (0-10)	Peak current and Pulse on Time had significant effect on both MRR and SR

Ishfaq *et al.* [47] examined process parameters of WEDM for machining of stainless clad steel. Machining of cladded material is very challenging as of their heterogeneous properties. WEDM of cladded steel provide better surface finish in comparison to other processes such as gas cutting and plasma arc cutting specifically used to cut these materials. However, the low cutting speed is a limitation of using WEDM. Experiments were designed based on Taguchi method to optimize cutting speed. It was seen that individual layer thickness of cladded material had crucial part to control the machining rate of WEDM. Babu *et al.* [48] investigated the parametric optimization of hypereutectic Al-Si alloy in WEDM. For experimentation, different input controlling factors were selected including pulse on time and variation of silicon percentage. GRA and principal component analysis were implemented for multi objective optimization. PCA was found better in performing the task in comparison to GRA.

Periyanan *et al.* [49] described the utilization of micro-WEDM in machining intricate and accurate 3D structural micro parts and micro tools. Taguchi was employed to optimize the input variables of micro-WEDM for enhancing MRR. Results showed that feed rate, capacitance and voltage were most affecting parameters on MRR.

Giridharan and Samuel [50] conducted the experiment on AISI 4340 steel in WEDT to improve its erosion rate. Newer method was adopted to measure the carter depth and better results were obtained. The stochastic erosion mechanism was investigated with SEM Table **3** shows a review on WEDM from various studies.

FUTURE PROSPECTS OF WEDM

Frequent wire ruptures make WEDM less efficient and degrade machining quality. Thus, reducing wire rupture for better dimensional accuracy and efficiency will remain a key research area.

Titanium alloys have vast application. Their machining is still a challenge as these alloys are hard to machine into intricate shapes. Process parameters of WEDM can be optimized to cut titanium alloys in order to reduce manufacturing cost and time.

Machining efficiency of WEDM for titanium alloys should be enhanced. However, existing wire materials have failed in these regards. Therefore, new wire electrode materials with integration of new core and coating material are required.

CONCLUSION

After a thorough study of WEDM literature, it can be concluded that titanium alloys are very useful in the area of biomedical, aerospace and micro tool manufacturing. These alloys can be machined with intricate shapes smoothly and inexpensively with WEDM. It has been observed that input variables have significant effect on performance measures, so it become of paramount importance to optimize the parameters to achieve high dimensional accuracy and high repeatability. Optimization techniques such as Taguchi, RSM, Fuzzy, algorithm and ANN based on mathematical models can be employed for such investigation for various titanium alloy machining in WEDM. Moreover, with the application of SEM, TEM and XRD, surface integrity, composition and defects in machined materials can be evaluated and studied.

CONSENT FOR PUBLICATION

Not applicable.

CONFLICT OF INTEREST

The authors confirm that this chapter content has no conflict of interest.

ACKNOWLEDGEMENTS

The authors would like to express their sincere thanks to the editor and anonymous reviewers for their time and valuable suggestions.

REFERENCES

[1]   C-T. Lin, "I-Fang Chung and Shih-Yu Huang, "Improvement of machining accuracy by fuzzy logic at corner parts for wire-EDM", *Fuzzy Sets Syst.,* no. 122, pp. 499-511, 2001.
[http://dx.doi.org/10.1016/S0165-0114(00)00034-8]

[2]   A.B. Puri, and B. Bhattacharyya, "An analysis and optimisation of the geometrical inaccuracy due to wire lag phenomenon in WEDM", *Int. J. Mach. Tools Manuf.,* no. 43, pp. 151-159, 2003.
[http://dx.doi.org/10.1016/S0890-6955(02)00158-X]

[3]   N. Tosun, and C. Cogun, "An investigation on wire wear in WEDM", *J. Mater. Process. Technol.,* no. 134, pp. 273-278, 2003.
[http://dx.doi.org/10.1016/S0924-0136(02)01045-2]

[4]   A.B. Puri, and B. Bhattacharyya, "Modelling and analysis of the wire-tool vibration in wire-cut EDM", *J. Mater. Process. Technol.,* vol. 141, pp. 295-301, 2003.
[http://dx.doi.org/10.1016/S0924-0136(03)00280-2]

[5]   D. Amin, and V. Mehta, "A Review Paper on Wire Electric Discharge Machining of Cryo treated Ti6Al4V", *International Journal for Innovative Research in Science & Technology,* vol. 2, no. 12, pp. 173-177, 2016.

[6]   K.P. Farnaz Nourbakhsh, "Wire electro-discharge machining of titanium alloy", *procedia CIRP,* no. 5, pp. 13-18, 2013.

[7]   Bassam Khan, Rahul Davis, and Abhishek Singh, *Effect of input variables and cryogenic treatment in wire electric discharge machining of Ti-6Al-4V alloy for biomedical applications.*
[http://dx.doi.org/10.1016/j.matpr.2019.09.226]

[8]   K.H. Ho, S.T. Newman, S. Rahimifard, and R.D. Allen, "State of the art in wire electrical discharge machining (WEDM)", *Int. J. Mach. Tools Manuf.,* no. 44, pp. 1247-1259, 2004.
[http://dx.doi.org/10.1016/j.ijmachtools.2004.04.017]

[9]   J. Wang, and B. Ravani, "Computer aided contouring operation for traveling wire electric discharge machining (EDM)", *Comput. Aided Des.,* no. 35, pp. 925-934, 2004.

[10]   S. Swarup, "Optimization of Process Parameters of Wire Electric Discharge Machining on AISI 4140 Using Taguchi Method and Grey Relational Analysis", *Materials Today: Proceedings,* no. 18, pp. 4261-4270, 2019.

[11]   Jaksan C Patel and Kalpesh D Maniya, "A Review on: Wire cut electrical discharge machining process

for metal matrix composite", *Procedia Manufacturing,* no. 20, pp. 253-258, 2018.

[12]   Ibrahem Maher, "Review of improvements in wire electrode properties for longer working time and utilization in wire EDM machining", *International Journal of Advance Manufacturing and Technology,* no. 76, pp. 329-351, 2015.

[13]   J. Kapoor, "High performance wire electrodes for wire electrical-discharge machining – a review", *Proceedings of the Institution of Mechanical Engineers, Part B: Journal of Engineering Manufacture,* 2012pp. 1-17

[14]   M. Parthiban, V. Krishnaraj, R. Sindhumathi, and J. Valentincic, "Investigation on manufacturing of microtools made of tungsten carbide using wire electric discharge grinding (WEDG)", *J. Braz. Soc. Mech. Sci. Eng.,* 2017.
[http://dx.doi.org/10.1007/s40430-017-0780-2]

[15]   T. Masuzawa, M. Fujino, and K. Kobayashi, "Wire Electro-Discharge Grinding for Micro-Machining", *Annals of the CIRP,* vol. 34, no. 1, pp. 431-434, 1985.
[http://dx.doi.org/10.1016/S0007-8506(07)61805-8]

[16]   Y. Zhu, T. Liang, L. Gu, and W. Zhao, "Machining of Micro Rotational Parts with Wire EDM Machine", *Procedia Manufacturing,* vol. 5, pp. 849-856, 2016.
[http://dx.doi.org/10.1016/j.promfg.2016.08.071]

[17]   M.Y. Ali, M.A. Rahman, and R. Nordin, "Micro Wire Electro Discharge Grinding: Optimization of Material Removal Rate and Surface Roughness", *IOP Conf. Series: Materials Science and Engineering,* 2017pp. 1-7
[http://dx.doi.org/10.1088/1757-899X/184/1/012032]

[18]   Abimannan Giridharan and G. L. Samuel, "Modeling and analysis of crater formation during wire electrical discharge turning (WEDT) process", *International Journal of Advance Manufacturing and Technology,* no. 77, pp. 1229-1247, 2015.

[19]   V.P. Srinivasan, and P.K. Palan, "Surface integrity, fatigue performance and dry sliding wear behaviour of Si3N4–TiN after wire-electro discharge machining", *Ceram. Int..*
[http://dx.doi.org/10.1016/j.ceramint.2020.01.082]

[20]   A.M. Takale, and N.K. Chougule, "Effect of wire electro discharge machining process parameters on surface integrity of Ti49.4Ni50.6 shape memory alloy for orthopedic implant application", *Mater. Sci. Eng. C,* vol. 97, pp. 264-274, 2019.
[http://dx.doi.org/10.1016/j.msec.2018.12.029] [PMID: 30678911]

[21]   B. Singh, and J.P. Misra, "Surface finish analysis of wire electric discharge machined specimens by RSM and ANN modeling", *Measurement,* no. 137, pp. 225-237, 2019.
[http://dx.doi.org/10.1016/j.measurement.2019.01.044]

[22]   A. Pramanik, and A.K. Basak, *Effect of wire electric discharge machining (EDM) parameters on fatigue life of Ti-6Al-4V alloy,* 2019.
[http://dx.doi.org/10.1016/j.ijfatigue.2019.105186]

[23]   V. Kaviman, and K. Soorya Prakash, "Influence of machining parameters on wire electrical discharge machining performance of reduced graphene oxide/magnesium composite and its surface integrity characteristics", *Composites Part B,* no. 167, pp. 621-630, 2019.

[24]   X. Chu, X. Zeng, W. Zhuang, W. Zhou, X. Quan, and T. Fub, "Vibration assisted high-speed wire electric discharge machining for machining surface microgrooves", *J. Manuf. Process.,* no. 44, pp. 418-426, 2019.
[http://dx.doi.org/10.1016/j.jmapro.2019.05.026]

[25]   K. Rajkumar, C. Balasubramaniyan, K.M. Nambiraj, and M. Rajesh, *A brass wire electric discharge machining variable optimization of aluminium–nano alumina composite.*
[http://dx.doi.org/10.1016/j.matpr.2019.11.252]

[26]   Guojun Zhang, Wenyuan Li, Yanming Zhang, Yu Huang, Zhen Zhang, and Zhi Chen, *Analysis and*

reduction of process energy consumption and thermal deformation in a micro-structure wire electrode electric discharge machining thin-wall component, 2020.
[http://dx.doi.org/10.1016/j.jclepro.2019.118763]

[27]   Neeraj Sharma and Kapil Gupta, "Wire Spark Erosion Machining of Ni rich NiTi Shape Memory Alloy for Bio-Medical Applications", In: *Procedia Manufacturing*, 2019, no. 35, pp. 401-406.

[28]   A. Smirnov, P. Peretyagin, and J.F. Bartolomé, "Wire electrical discharge machining of 3Y-TZP/Ta ceramic-metal composites", *J. Alloys Compd.,* 2018.
[http://dx.doi.org/10.1016/j.jallcom.2017.12.221]

[29]   C. Zhou, X. Wu, and Y. Lu, "Wen Wu, Hang Zhao and Liejun Li, "Fabrication of hydrophobic Ti3SiC2 surface with micro-grooved structures by wire electrical discharge machining", *Ceram. Int.,* pp. 1-8, 2018.
[http://dx.doi.org/10.1016/j.ceramint.2018.07.032]

[30]   A. Pramanik, and A.K. Basak, "Sustainability in wire electrical discharge machining of titanium alloy: understanding wire rupture", *J. Clean. Prod.,* pp. 1-16, 2018.
[http://dx.doi.org/10.1016/j.jclepro.2018.07.045]

[31]   K. Titus Thankachan, "Soorya Prakash, R Malini, S Ramu, Prabhu Sundararaj, Sivakumar Rajandran, Devaraj Rammasamy and Sathiskumar Jothi, "Prediction of Surface roughness and Material removal rate in Wire Electrical Discharge Machining on Aluminum Based Alloys/Composites using Taguchi Coupled Grey Relational Analysis and Artificial Neural Networks", *Appl. Surf. Sci.,* 2018.
[http://dx.doi.org/10.1016/j.apsusc.2018.06.117]

[32]   Somvir Singh Nain, Dixit Garg, and Sanjeev Kumar, *Investigation for obtaining the optimal solution for improving the performance of WEDM of super alloy Udimet-L605 using particle swarm optimization,* 2018.
[http://dx.doi.org/10.1016/j.jestch.2018.03.005]

[33]   Tina Chaudhary, Arshad Noor Siddiquee, and Arindam Kumar Chanda, "Effect of wire tension on different output responses during wire electric discharge machining on AISI 304 stainless steel", , pp. 1-13, 2018.
[http://dx.doi.org/10.1016/j.dt.2018.11.003]

[34]   V.N. Gaitonde, M. Manjaiah, S. Maradi, S.R. Karnik, P.M. Petkar, and J. Paulo Davim, *Multiresponse optimization in wire electric discharge machining (WEDM) of HCHCr steel by integrating response surface methodology (RSM) with differential evolution.* Computational Methods and Production Engineering: DE, 2017, pp. 199-221.
[http://dx.doi.org/10.1016/B978-0-85709-481-0.00007-0]

[35]   A. Goyal, ""Investigation of material removal rate and surface roughness during wire electrical discharge machining (WEDM) of Inconel 625 super alloy by cryogenic treated tool electrode," Journal of King Saud University –", *Science,* no. 29, pp. 528-535, 2017.

[36]   K. Mouralova, J. Kovar, L. Klakurkova, T. Prokes, and M. Horynova, "Comparison of morphology and topography of surfaces of WEDM machined structural materials", *Measurement.*
[http://dx.doi.org/10.1016/j.measurement.2017.03.009]

[37]   K. Siva Prasad Arikatla, "Parametric Optimization in Wire Electrical Discharge Machining of Titanium Alloy Using Response Surface Methodology", *Materials Today: Proceedings,* no. 4, pp. 1434-1441, 2017.

[38]   T.U. Siddiqui, and J. Ramkumar, ""Micro-wire electric discharge machining of Mg alloy used in biodegradable orthopaedic implants," Materials Today", *Proceedings,* vol. 4, pp. 10273-10277, 2017.

[39]   J.F. Liu, Y.B. Guo, T.M. Butler, and M.L. Weaver, "Crystallography, compositions, and properties of white layer by wire electrical discharge machining of nitinol shape memory alloy", *Mater. Des.,* vol. 109, pp. 1-9, 2016.
[http://dx.doi.org/10.1016/j.matdes.2016.07.063]

[40]   M. Manjaiah, S. Narendranath, S. Basavarajappa, and V.N. Gaitonde, "Effect of electrode material in wire electro discharge machining characteristics of Ti50Ni50–xCuxshape memory alloy", *Precis. Eng.*, 2015.
[http://dx.doi.org/10.1016/j.precisioneng.2015.01.008]

[41]   Y. Sun, Y.D. Gong, X.L. Wen, G.Q. Yin, and F.T. Meng, "Micro milling characteristics of LS-WEDM fabricated helical and corrugated micro end mill", *Int. J. Mech. Sci.*, 2019.
[http://dx.doi.org/10.1016/j.ijmecsci.2019.105277]

[42]   Saeid Shakeri, Aazam Ghassemi, Mohsen Hassani, and Alireza Hajian, *Investigation of material removal rate and surface roughness in wire electrical discharge machining process for cementation alloy steel using artificial neural network,* 2016.
[http://dx.doi.org/10.1007/s00170-015-7349-y]

[43]   A. Goyal, and U.R. Huzef, "Rahman and S.A.C. Ghani, "Experimental investigation & optimisation of wire electrical discharge machining process parameters for Ni49Ti51 shape memory alloy," Journal of King Saud University -", *Eng. Sci.*, 2020.
[http://dx.doi.org/10.1016/j.jksues.2020.01.003]

[44]   S. Suthan, and K. Muralidharan, "Experimental analysis optimization of process parameters of wire edm on stainless steel316L", *International Research Journal of Engineering and Technology,* vol. 05, no. 06, pp. 1258-1263, 2018.

[45]   Y.S. Liao, J.T. Huang, and Y.H. Chen, "A study to achieve a fine surface finish in Wire-EDM", *J. Mater. Process. Technol.,* no. 149, pp. 165-171, 2004.
[http://dx.doi.org/10.1016/j.jmatprotec.2003.10.034]

[46]   K.R. Padmavathi, "2 S. Devaraj, 3 I. John Solomon, 4 A. Premkumar, "Influence of Process Parameters on Wire EDM Process for AISI 316 Stainless Steel", *Int. J. Eng. Res. Technol. (Ahmedabad),* vol. 06, no. 02, 2018.
[http://dx.doi.org/10.17577/IJERTCON067]

[47]   K. Ishfaq, N.A. Mufti, M.P. Mughal, M.Q. Saleem, and N. Ahmed, "Investigation of wire electric discharge machining of stainless-clad steel for optimization of cutting speed", *Int. J. Adv. Manuf. Technol.,* 2018.
[http://dx.doi.org/10.1007/s00170-018-1630-9]

[48]   D.M. Babu, S.V. Kiran, and P.R. Vundavilli, "Experimental investigations and multi-response optimisation of wire electric discharge machining of hypereutectic Al-Si alloys", *Int. J. Manuf. Res.,* vol. 11, no. 03, pp. 221-237, 2016.
[http://dx.doi.org/10.1504/IJMR.2016.079460]

[49]   P.R. Periyanan, U. Natarajan, and A. Elango, "Optimization of parameters on material removal rate in micro-WEDG process", *Int. J. Eng. Sci. Technol.,* vol. 3, no. 9, pp. 66-76, 2011.

[50]   A. Giridharan, and G.L. Samuel, "Investigation into erosion rate of AISI 4340 steel during wire electrical discharge turning process", *Mach. Sci. Technol.,* vol. 22, no. 2, pp. 287-298, 2018.
[http://dx.doi.org/10.1080/10910344.2017.1365890]

CHAPTER 8

Recent Developments in EDM and its Types

Anupam Thakur[*] and **Suneev Anil Bansal**

Department of Mechanical Engineering, Maharaja Agrasen University, Baddi, Himachal Pradesh, India

Abstract: Electric discharge machining has been the one of the advanced machining processes that deals with the machining of intricate and complex shapes on a workpiece considering various parameters like surface roughness and higher accuracy. A lot of research has been carried out in the EDM area for developing various models and selecting the best optimum condition for machining using conventional or wire EDM. Some results regarding submerged and dry EDM processing have also been discussed. Along with this, a recent development of smart EDM is also discussed for predicting the future capabilities of an EDM machine. In this article a light has being thrown on all the recent developments being done on the basis of machining capabilities, parameter, along with the characteristics of machining surface and various wire materials and the effect on the EDM wire during machining and the overall performance.

Keywords: Composites, Electric discharge machine, Materials, MRR, Process parameters.

INTRODUCTION

Electric discharge machine facilitates many intricate complex shapes being produced without the physical contact between the workpiece and the tool in form of electrode. The process of machining is carried out by continuous spark erosion between the electrode and workpiece interface (Fig. **1**). This process is carried out by developing an appropriate voltage between the workpiece and the tool which leads to the dielectric medium breakdown as a result of strong electrostatic field. It is an advanced manufacturing technique generally used in die manufacturing mould making, automotive manufacturing, and medical instruments *etc.* Most fragile objects, cannot be machined easily by other machines, can easily be machined without any risk factor. It has been seen that a no toxic gas fumes and contaminated dielectric are been produced.

---

[*] **Corresponding author Anupam Thakur**: Department of Mechanical Engineering, Maharaja Agrasen University, Baddi, Himachal Pradesh, India; E-mail:anupamthakur@hotmail.com

Wire EDM is also somehow similar to a conventional EDM as it implements the electrode in the form of wire that is used for spark erosion.

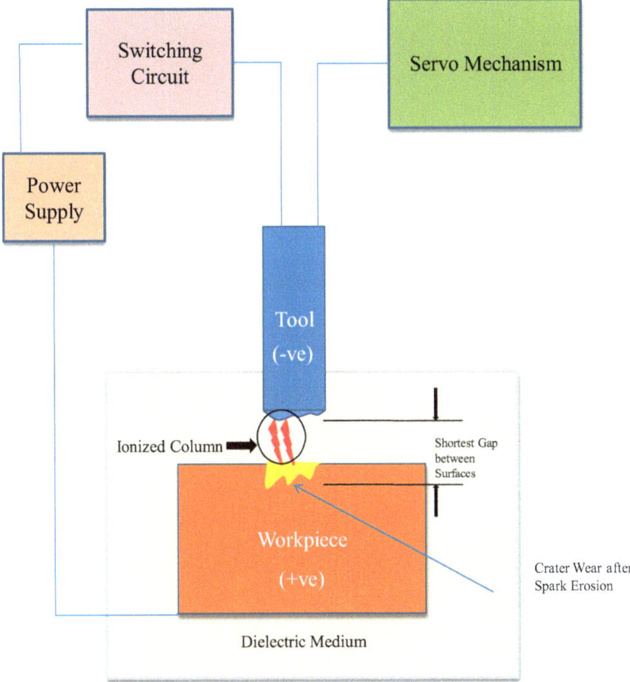

Fig. (1). Layout of EDM Mechanism.

This further gave more scope for producing intricate and fine shapes having high surface finish and accuracy to be produced using WEDM. Some researches depict the creation of ionized column at the shortest distance between the tool and workpiece [1]. With the advancement, complex and fine design, the wire EDM has come into light which can give a new scope of machining with high level of accuracy and finishing. It uses electricity as the source conducting through a thin copper or brass wire acting as an electrode. When this wire comes closer with the workpiece, the electric spark at the nearest section starts thus leading to spark erosion of the metal by melting and vaporization of microscopic particles. During this process, wire also gets eroded on a minute sale thus leading to a form of tool wear. To allow an equal and precise dimension of the cutting area, the machine changes the old wire with new wire section by automatic mechanism present on the machine as the machining advances and cutting is performed. The whole process is carried out in a dielectric medium of deionized water to facilitate evenly and continuous cooling of the work area with removal of eroded particles produced during machining.

CURRENT STATE OF EDM AND EFFECT OF PULSE NATURE ON EDM PROCESS

Due to non-linearity in EDM, a lot of improvements are required in the EDM machines. Several researches have highlighted the aspects that could be improved in order to improve the current process (Table **1**). During the last 20 years, a lot of research has been done starting from the first-time developments alongwith its progress with time [2]. It has given various control and results for improvement in the EDM machine with optimization of the process parameters [3]. Keeping in view of optimizing, the input parameters was stated in the study. Some study also emphasized on the new technological developments to improve the quality of machined surface. It was also found that the large number of studies focused on powder mix EDM to study surface and its modification [4]. Studies have also been done on the DC electric pulses that are been used to increase the efficiency of the machine. It has been found that if the improvement is being made in the pulse generator using RC type, then a lot better results can be seen, comprising of high surface quality and improved MRR using tungsten carbide as workpiece [5] (Fig. **2**).

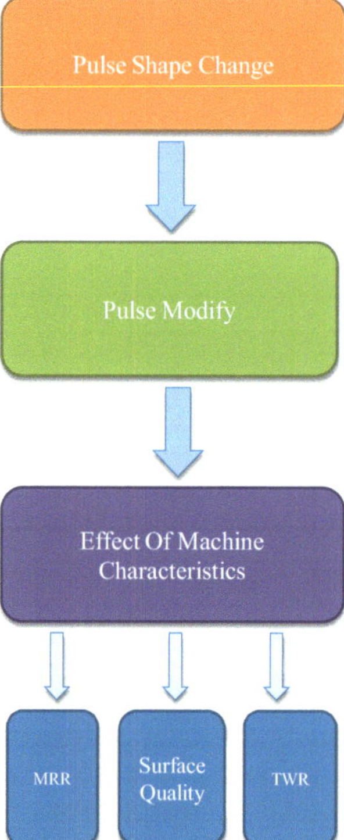

Fig. (2). Pulse shape effect on EDM characteristics.

Another study shows that if a pulse of 1Mhz from a modified transistor is used, instead of a RC type generator, then it produces higher MRR with a three times higher machining speed [6]. Han *et al.* conducted a study in which they designed a new transistor pulse generator for pulse generation which was able to produce 24 times greater MRR [7]. A research was also carried out, in which, an effect of EDM capacitance was seen on the impedance thus leading to high machining properties [8]. A lower energy, dispensed at the tool and workpiece interface, was also implemented during the surface finishing time to get higher surface finish using some semiconductor-based switching circuit [9].

Some experiments showed that the shape change of pulse generator, being used for machining, had a significant effect on the various machining characteristics. It was shown that energy density had a large effect on the tool wear rate (TWR) [10]. Short pulse duration often was used for high precision machining. The surface morphology was found to be of higher value when using a uniform discharge of energy. It was also found that less arcing effect occurred if uniform shape of pulse for discharge machining was used [11].

Table 1. Various Reviews for Wire EDM.

S.No.	Author	Work Material	Input Parameters	Result
1	Bhatia *et al.* [12]	High carbon high chromium steel	Pulse on time(125-131μs), pulse off time(34-48 μs), wire tension(1-3), max. current(210-230A).	Effect of pulse off time was found significant on SR. Model developed had an error of 3.93%.
2	Tilekar *et al.* [13]	Aluminium and Mild Steel	Pulse on time(25-39 μs), pulse off time(6-10μs), current(1-3A), wire feed rate(75-85mm/s).	SR was effected by pulse on time and current. For Kerf width Wire feed had significant effect.
3	Brajesh *et al.* [14]	AISI D3 Steel	Pulse on time(18-24 μs), pulse off time(51-59μs), peak current(180-200A), wire feed rate (4-6mm/min).	SR was effected
4	Selvakumar *et al.* [15]	5083 Aluminium Alloy	Pulse on time(0.5-0.9 μs), pulse off time(14-38μs), peak current(20-100A), wire tension (420-660g).	Cutting speed was independent of the wire tension, surface roughness (Ra) was independent of pulse off time and tension in wire.
5	Azhiri *et al.* [16]	Al/SiC	Pulse on time(106-126 μs), pulse off time(60-40μs), discharge current(70-230A),gap voltage(60-20V).	Cutting velocity and surface roughness were effected by pulse on time and current. Grey rational analysis were performed to optimize these two also.

(Table 1) cont.....

S.No.	Author	Work Material	Input Parameters	Result
6	Srivastava *et al.* [17]	Al2024 reinforced with SiC	Current(2-4A), Pulse on time(15-25 µs), Reinforcement Percentage (2-4%)	SR increased with current, pulse on time & reinforced percentage. MRR varies linearly with pulse on time and current and inversely with the Reinforcement percentage.
7	Ikram *et al.* [18]	Tool steel D2	Wire feed velocity (5-15m/s), pulse on-time(3-5 µs), pulse off-time(2--30 µs), dielectric pressure(4-10 kg/cm^2),wire tension(800-2000g),open voltage(80-110V) servo voltage(40-60V) and material thickness(12.7- 25.4)	Pulse on time is significant for SR, kerf width and MRR.
8	Bobbili *et al.* [19]	High strength Armor steel	Pulse on time (0.6-1.6µs), Pulse off time(14-46 µs), Wire feed(2-4m/min.), flushing pressure(4-8Kg/cm^2), spark voltage(15-25V), wire tension(4-8Kg-f)	A regression model was developed for the various response parameters and the result show that Spark voltage, Pulse on/off time has high level of significance for MRR and Surface roughness.
9	Kumar *et al.* [20]	Pure Titanium	Pulse on time (0.7-1.1µs), Pulse off time(17-38 µs), Wire feed(4-10m/min.), peak current(120-200A), spark voltage(40-60V), wire tension(500-1400g)	A semi empirical model was developed, comparison between measured and predicted MRR with an error of 6-32-6.95%
10	Zhang *et al.* [21]	SKD11	Pulse on time (1-9µs), Pulse off time(10-70 µs), Wire feed(3-11m/min.), current(1-5A),Tracking Coefficient (40-80)	Optimal conditions for response parameters high MRR and low roughness were modeled.
11	Garg *et al.* [22]	Titanium 6-2-4-2 Alloy.	T_{on}(0.7-.0µs),T_{off}(22-38), IP(140-200A), WF(6-10m/min.), WT(500-1000g)	CV=8.1235mm/min. and SR= 1.2549µm optimized for best condition for this MMC

MRR= Material Removal Rate, SR,Ra= Surface Roughness, TWR= Tool Wear Rate, Ton= Pulse on time, Toff= Pulse off Time, WT=Wire tension, WF=Wire feed, CV= Cutting Velocity.

EFFECT OF WIRE POSITIONING

It has been found that the dimensional consistency of the wire used in WEDM plays a very important role in the accuracy and preciseness of machining. Also, with it, the accuracy of the worktable plays a very important role. Due to the presence of various other factors, this precision cannot be reached theoretically. As the controlling wire control during the machining area is very difficult (Fig. **3**). Controlling, back and forth motion of the wire during sparking and cutting, is very difficult to hold at a precise position. Also, the section of wire, free of holding, is susceptible to vibration frequency. This is prevented by computer control that monitors the gap between the wire and the workpiece. But there are some other factors and reasons due to which some of the vibration cannot be controlled, no matter how better the positioning system may be. This results in the gap change between the electrode wire and the workpiece in spark area which results in a continuous change in the breakdown voltage and discharge energy. It has been studied that this oscillation of wire, due to internal and external vibration, the wire can result in an oscillation of up to 10μm [23]. So machines need to be made as rigid as possible to prevent this vibration which ultimately would result in a high surface finish of the workpiece.

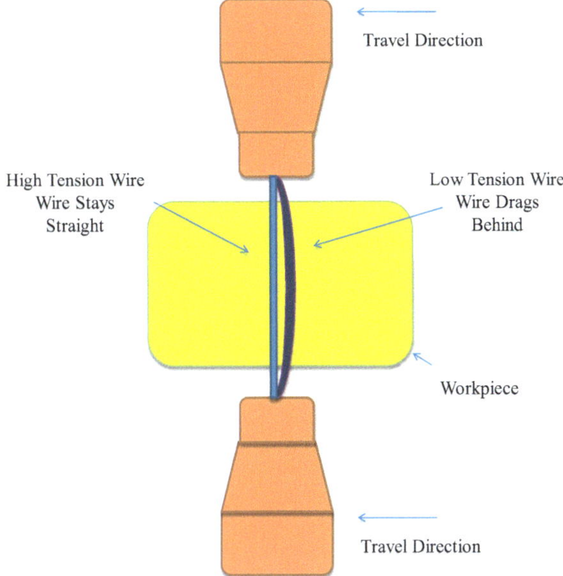

Fig. (3). Wire drag and Tension produced during machining in WEDM [25].

EFFECT OF WIRE TENSION

To prevent the vibration, in the wire during machining, the two ends of the wire are kept under tension thus minimizing vibrations. This helps in keeping the wire in straight form and maintaining a good cutting. This also leads to an equal gap of cut during the travel of wire. Still there can be the breakage of the wire due to excessive stresses produced in the wire. It acts as a limiting condition to the tension applicable in wire (Fig. **4**) [23, 24].

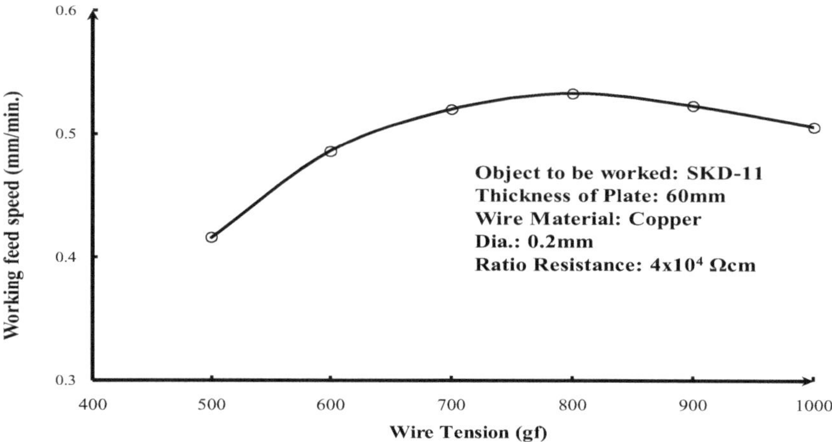

Fig. (4). Graph showing the cutting speed variation with wire tension [23].

EFFECT OF WIRE MATERIAL AND COMPOSITION

As it has been seen in the previous section that the wire vibration and its rigidity during the machining process plays a very important role. Thus, material selection for the wire should be made on the basis of having high endurance in terms of tensile strength and toughness to absorb all the vibrations and stresses produced with minimum vibration. It has been studied that with the rise of Zn percentage in the wire composition the machining speed increases with lower servo voltage requirement thus leading to lower chances of short circuit during machining [26]. A lot have researched have been done to maximize MRR with least TWR and SR. Few of the important wire material used are discussed below.

Copper

Copper was the one of the first materials to be used as an electrode for machining. But due to the physical limitation in the properties like strength and low vapour pressure it puts some limitation to it use as wire in WEDM [25].

Brass

As an alternative to Cu, Br plays a very significant role in WEDM having good performance results. It is an alloy of Cu and Zn containing both the elements up to around 63% and 37% respectively. Zn is added to exhibit high tensile strength, high vapor pressure and low melting point [25, 26].

Coated Wires

It has been found that the electrode coating of wires, having less vapourization temperature, possesses fewer chances to experience thermal shock. It includes a wire having high conductivity core with a coating of an alloy having low melting point and finally a coating of alloy having high strength, thus facilitating high machining speed. A lot of researches of such coated electrode wire have been carried out having high machining capabilities with better flushablity producing cuts with least inaccuracy and surface roughness. But the only disadvantage is the high coat along with the chances of contamination of dielectric used creating environmental limitations [27].

Diffusion Annealed Wires

Due to such good characteristics of the zinc coating it can be thought that a pure zinc coated wire would produce a high machining result. Zn offers low melting point, combined with good flushablity. Thus, it can be stated that a diffused coating having high or pure concentrations of Zn on the top layer could produce encouraging results. To produce such diffused wires, the annealing of the encasing material is done that allow the even distribution of the alloy on top layer. By this one can produce a wire having high resistance to wear as compared to the conventional electrode material. The resulting EDM wire resulted in a quite significant increase in production rate as compared to the brass electrode wire [28].

EFFECT OF FLUSHING AND DIELECTRIC FLUID

Dielectric is used in EDM in the various forms such as full immersion, flushing with jet or spray. In Fig. (5) the effect of flushing pressure on the machining speed (Vs) and Surface roughness (Ra) is being plotted.

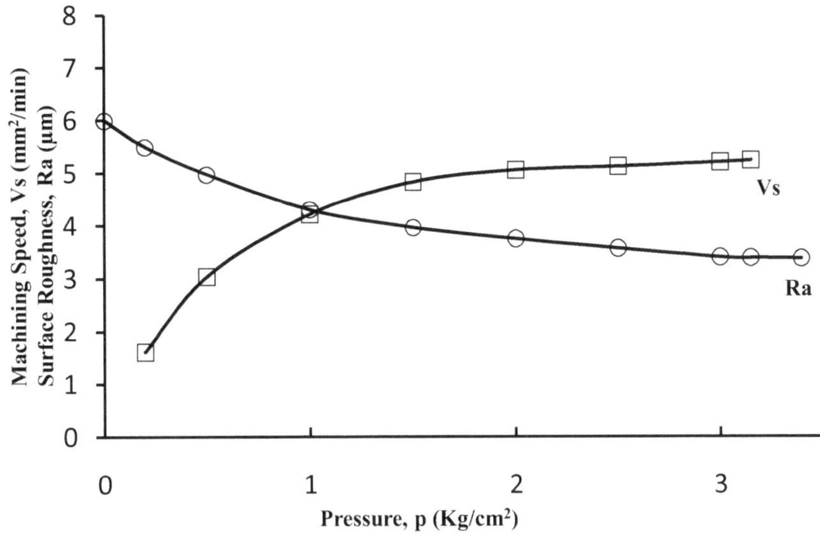

Fig. (5). Effect of Pressure on Vs and Ra [23].

The pressure flow facilitates the continuous removal of the eroded particles created during sparking and it should be less when approaching the finishing time. It is also seen there a certain limit out of which the adequate machining cannot be done, with increasing flush pressure upto 1 Kg/cm^2 higher machining speeds are possible, beyond this there is decrease in machining speed rate. High temperature ranges are seen along spark discharge when flushing pressure is less than 0.3 kg/cm^2.

The dielectric fluid other than flushing out the eroded particles performs various functions such as cooling the workpiece and tool during machining. It also helps to concentrate the spark energy on the required smaller section of machining and gap insulation before accumulation of large amount of energy.

SMART WIRE EDM

As we know that industry has been changing and it has been introduced with new revolution called as 'Industry 4.0'. Studies prove that this could produce much better results by cutting off the various unwanted cost by avoiding downtime [29]. It has been seen that with the current requirement of micro machining some intelligence factors need to be introduced which could help do such work with high level of accuracy and precision. A lot of funds have been seen to be invested in such projects as compared to conventional EDM projects [30]. In this it has been emphasized to introduce a new intelligence factor into the machining era which could optimize various control parameters and exclude environmental hazards. These not only restrict to this but also have the capability to self-investigate and monitor the various tasks being done and do any appropriate changes to get the best machining characteristics. Such implication is being seen in recent year in a research done by Walder *et al.* [31]. As WEDM are equipped with no of sensors and devices to monitor various things such as replaceable part as their service hour are completed a lot of manufacturing cost can be cut as compared to a conventional EDM where we replace some parts much before there service life gets finished to avoid downtime but increasing manufacturing cost at the same time. No deep study has been done till to optimize such and build a prediction model as it requires a lot of thorough study in field of machining along with its chemical study. An effort has been made to build such model by Walder *et al.* This has various advantages like real time data collection and updating of the prediction model to get a precise model as the machining is being carried out. Recent applications of WEDM are seen in the materials like Inconel, work steel K110 and other hard metal being used in various high-grade industries. Study was focused on the four parameters *i.e.* EDM wire, Main Filter, Resin for deionization, Contacts for current supply to wire as these were having major down-time/replacement cost.

The dielectric fluid plays a very important role as it cleans the machining surface continuously, maintains the adequate electric conductivity while machining. Consumable items like filters and dielectric depend a lot on the material to be machined and various control parameters. Sensors are installed on the machine to have a continuous monitoring of these equipment. Dielectric used was merely deionized water having a conductivity merely of 15 µ-siemens/cm.

The dielectric used here plays a major role of maintaining the temperature of 20°C along with cleaning and maintain the deionization gap between the electrode and the workpiece. For these two reservoirs are used which have filters installed in them and maintain the pressure of the dielectric and continuously clean the dielectric. In Fig. (6) we can see the layout of dielectric circuit being built to stabilize the dielectric conductivity, as it changes with the introduction of new fine particles produced while machining. So, inspite of having a large no of sensors been prior installed a new deionization system is being used to lower this increased conductivity and stabilize it.

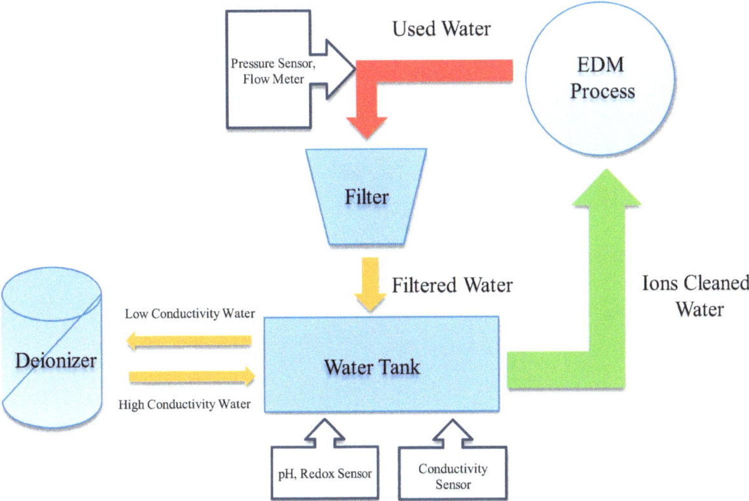

Fig. (6). Layout of Dielectric Circuit [31].

This consists of a pressure sensor, conductivity sensor and a combination of pH and redox sensor in the places. Also, a monitor on filter pressure drop was maintained using pressure sensor at the filter entrance. It was set for a pressure vale of upto 2.8 bar, upon crossing which the machine stops automatically. Upon machining the given material a pressure drop chart for each was obtained and the weight of the eroded particles from the resin was calculated for each setup which shows that the machining of about 304 effective hours a total of 4kg of WC was being removed by machining of workpiece and wire of around 7.7 kg was eroded during machining. Compared to the amount of debris collected in resin a total of 5.4 Kg was found to be absorbed as in the form of ions during machining. Finally, a model was created for predicting the pressure drop. Contact wear was also found to be of significant rate as it continuously rubs against the wire of EDM. A good relation between the literature model and the developed model was found. Finally, it was concluded that the resin, filters, contacts and wire used in WEDM turn to be most cost effective if monitored precisely using appropriate sensors and prediction model.

CONCLUDING REMARKS

A lot of research has been done for the various material including the composites also considering various control parameters such as pulse on / off time, wire feed velocity, wire tension, peak current and voltage, but it can be seen from the review been done so far that a small study is done on the effect of wire material, so thorough studies should be done for finding the significance of it on the machining characteristics. It is also being seen that the wire diameter may play an important role in the machining characteristics and WWR. This also needs to be studied for their significance on MRR, SR and CV. Various other parameters such as filter saturation was found to be quiet different as compared to the literature model, so a detailed model need to be model for predicting it using materials like aluminium. We also see that as assumption of missing mass to be consumed in the form of ions has been made. Some research needs to be done to verify this and elaborate it thoroughly.

CONSENT FOR PUBLICATION

Not applicable.

CONFLICT OF INTEREST

The authors confirm that this chapter content has no conflict of interest.

ACKNOWLEDGEMENTS

The authors would like to express their sincere thanks to the editor and anonymous reviewers for their time and valuable suggestions.

REFERENCES

[1]   E.C. Jameson, *Electrical Discharge Machining.* 1ˢᵗ ed. Society of Manufacturing Engineers: Michigan, 2001.

[2]   K.H. Ho, and S.T. Newman, "State of the art electrical discharge machining (EDM)", *Int. J. Mach. Tools Manuf.,* vol. 43, pp. 1287-1300, 2003.

[3]   N.M. Abbas, D.G. Solomon, and M. Fuad Bahari, "A review on current research trends in electrical discharge machining", *Int. J. Mach. Tools Manuf.,* vol. 47, pp. 1214-1228, 2007.

[4]   S. Kumar, R. Singh, T.P. Singh, and B.L. Sethi, "Surface modification by electrical discharge machining: A Review", *J. Mater. Process. Technol.,* vol. 209, pp. 3675-3687, 2009.

[5]   M.P. Jahan, Y.S. Wong, and M. Rahman, "A study on the fine-finish die-sinking micro EDM of tungsten carbide using different electrode materials", *J. Mater. Process. Technol.,* vol. 209, pp. 3956-3967, 2009.

[6]   F. Han, L. Chen, D. Yu, and X. Zhou, "Basic study on pulse generator for micro EDM", *Int. J. Adv. Manuf. Technol.,* vol. 33, pp. 474-479, 2007.

[7]   F. Han, S. Wachi, and M. Kunieda, "Improvement of machining characteristics of micro-EDM using transistor type iso pulse generator and servo feed control", *Precis. Eng.,* vol. 28, pp. 378-385, 2004.

[8]   R. Casanueva, F.J. Azcondo, and S. Bracho, "Series-parallel resonant converter for an EDM power supply", *Journal of Materials Processing Technology,* vol. 149, 2004.

[9]   T. Muthuramalingam, and B. Mohan, "Design and fabrication of control system based iso pulse generator for electrical discharge machining", *International Journal of Mechatronics and Manufacturing Systems,* vol. 6, pp. 133-143, 2013.

[10]   T. Muthuramalingam, and B. Mohan, "Influence of discharge current pulse on machinability in electrical discharge machining", *Mater. Manuf. Process.,* vol. 28, pp. 375-380, 2013.

[11]   T. Muthuramalingam, and B. Mohan, "Experimental investigation of iso energy pulse generator on performance measures in EDM", *Mater. Manuf. Process.,* vol. 28, pp. 1137-1142, 2013.

[12]   A. Bhatia, S. Kumar, and P. Kumar, "A Study to achieve minimum surface roughness in Wire EDM", *Procedia Materials Science,* vol. 5, pp. 2560-2566, 2014.

[13]   S. Tilekar, and S.S. Das, "P.K Patowari, "Process Parameter Optimization of Wire EDM on Aluminum and Mild Steel using Taguchi Method", *Procedia Materials Science,* vol. 5, pp. 2577-2584, 2014.

[14]   B.K. Lodhi, and S. Agarwal, "Optimization of machining parameters in WEDM of AISI D3 Steel using Taguchi Technique", *Procedia CIRP,* vol. 14, pp. 194-199, 2014.

[15]   G. Selvakumar, G. Sornalatha, S. Sarkar, and S. Mitra, "Experimental investigation and multi-objective optimization of wire electrical discharge machining (WEDM) of 5083 aluminum alloy", *Trans. Nonferrous Met. Soc. China,* vol. 24, pp. 373-379, 2014.

[16]   R. Bagherian Azhiri, R. Teimouri, M. Ghasemi Baboly, and Z. Leseman, "Application of Taguchi, ANFIS and grey relational analysis for studying, modeling and optimization of wire EDM process while using gaseous media", *Int. J. Adv. Manuf. Technol.,* vol. 71, pp. 279-295, 2014.

[17]   A. Srivastava, A.R. Dixit, and S. Tiwari, "Experimental Investigation of Wire EDM Process Parameteres on Aluminum Metal Matrix Composite Al2024/SiC", *International Journal of Advance Research and Innovation,* vol. 2, pp. 511-515, 2014.

[18]   A. Ikram, N.A. Mufti, M.Q. Saleem, and A.R. Khan, "Parametric optimization for surface roughness, kerf and MRR in wire electrical discharge machining (WEDM) using Taguchi design of experiment", *J. Mech. Sci. Technol.,* vol. 27, no. 7, pp. 2133-2141, 2013.

[19]   V. Ravindranadh Bobbili, "Madhu, and A. K Gogia, "Effect of Wire-EDM Machining Parameters on Surface Roughness and Material Removal Rate of High Strength Armor Steel", *Mater. Manuf. Process.,* vol. 28, pp. 364-368, 2013.

[20]   A. Kumar, V. Kumar, and J. Kumar, "Semi-empirical Model On MRR And Overcut In WEDM ProcessOf Pure Titanium Using Multi- objective Desirability Approach", *J. Braz. Soc. Mech. Sci. Eng.,* vol. 37, pp. 689-721, 2015.

[21]   G. Zhang, Z. Zhang, J. Guo, W. Ming, M. Li, and Y. Huang, "Modeling And Optimization Of Medium Speed WEDM Process Parameters For Machining SKD11", *Mater. Manuf. Process.,* vol. 28, pp. 1124-1132, 2013.

[22]   M.P. Garg, A. Jain, and G. Bhushan, "Multi-objective Optimization of Process Parameters in Wire Electric Discharge Machining of Ti- 6-2-4-2 Alloy", *Arab. J. Sci. Eng.,* vol. 39, pp. 1465-1476, 2014.

[23]   A. Umare, S. Parchand, and P. Shingare, "Wire cut EDM Process- A review", *International Journal of*

Advance Research and Innovative Ideas in Education, vol. 3, no. 6, pp. 294-398, 2017.].

[24]   "Rishavraj Singh, "Review on Effects of Process Parameters in Wire Cut EDM and Wire Electrode Development", *International Journal for Innovative Research in Science & Technology,* vol. 2, no. 11, pp. 701-706, 2016.

[25]   D. Ghodsiyeh, A. Golshan, and J.A. Shirvanehdeh, "Review on Current Research Trends in Wire Electrical Discharge Machining", *Indian J. Sci. Technol.,* vol. 6, pp. 4128-4140, 2013.

[26]   J. Kapoor, Sehijpal Singh, and Jaimal Singh Khamba, "Recent Developments in Wire Electrodes for High Performance WEDM", *Proceedings of the World Congress on Engineering,* vol. 2, 2010.

[27]   M. Singh, H. Lal, and R. Singh, "Recent Developments in Wire EDM: A Review", *International Journal of Research in Mechanical Engineering & Technology,* vol. 3, pp. 150-152, 2013.

[28]   M.T. Antar, S.L. Soo, D.K. Aspinwall, D. Jones, and R. Perez, ""Productivity and workpiece surface integrity when WEDM aerospace alloys using coated wires," 1st CIRP Conference on Surface Integrity (CSI)", *Procedia Eng.,* vol. 19, pp. 3-8, 2011.

[29]   H. Kagermann, *Umsetzungsempfehlungen für Industrie 4.0,* p. 107.https://www.bmbf.de/files/ Umsetzungsempfehlungen_Industrie4_0.pdf

[30]   Q. Yi, *Micro-manufacturing research: drivers and latest developments,* 2015.https://strathprints.strath.ac.uk/id/eprint/59119

[31]   G. Wälder, D. Fulliquet, N. Foukia, F. Jaquenod, M. Lauria, R. Rozsnyo, B. Lavazais, and R. Perez, v "Smart Wire EDM machine, 19th CIRP Conference on Electro Physical and Chemical Machining", *Procedia CIRP,* vol. 68, pp. 109-114, 2018.

CHAPTER 9

Artificial Intelligence and Robotics in The Manufacturing Industry: Opportunities and Challenges

Hitesh Pahuja[*], **PK Khosla** and **Balwinder Singh**

Centre for Development of Advanced Computing, Mohali, Punjab, India

Abstract: Whatever is being done by humans today, will be done by Artificial Intelligence (AI) based systems, but at computer speed, scale and scope. AI is making its path towards the transformation of the manufacturing industry and improving the manufacturing processes. According to the panel of Forbes technology council members, 13 different sectors, including the manufacturing industry will be revolutionized by artificial intelligence. AI and robotics can transform the complicated task into consistent along with precise solutions. With the increasing operation of the industrial robot have drastically marked an increased deployment of AI in manufacturing processes. Artificial Intelligence (AI) is widely applied and is gradually approaching in the manufacturing segment, easing the automation with learning and error-free decision-making using learning algorithms. AI-driven systems are placing a simpler route to the future manufacturing industry by producing a lot of benefits by presenting innovative prospects, boosting production of goods with improving efficiencies, by implementing general AI to make machine interaction closer to human interaction. This technology enables the manufacturer to overcome various challenges that have remained in the manufacturing sector such as expertise shortage, poor decision making, problems associated with integration, and overloaded data. Hence, the implementation of AI in the manufacturing sector allows companies to fully renovate their procedures. It allows smart AI-robots/machines to collect and extract data, acknowledge various patterns, learn from that patterns and adapt to new environments through a machine or deep learning based on the complexity of tasks assigned to them. Therefore, the deployment of AI and robots in the industry will transform the mass-production along with the error-free decision-making capability. AI-based Robots in Industry 4.0 can do persistent actions such as from designing the manufacturing model, developing expertise, developing new automation solutions, eliminating human error along with providing exceptional quality in products. With the implementation of AI, robots will replace manpower and carry out hazardous activities. As a result, the number of accidents will drop across all sections in the manufacturing sector.

---

[*] **Corresponding author Hitesh Pahuja**: Centre for Development of Advanced Computing, Mohali, Punjab, India; E-mail: hitesh@cdac.in

AI & robotics technology will free the manufacturer issues related to untrained workers. Hence, entrepreneurs can concentrate on further modernization of industries and scaling their industry to greater heights. Apart from the numerous opportunities in adopting emerging technologies like AI, there are some challenges too for industrialists.The AI and robotics technology into the manufacturing industry would necessitate a huge capital investment for startups and existing manufacturers. But the return on investment will be high because smart machines will begin to take care of daily activities, companies will incur a considerably lower operating cost. AI's proponents claim that technology is only an evolutionary form of automation, a predictable result of the Fourth Industrial Revolution. AI may be efficient at creating things, improving them, and making them cheaper. But there is no replacement for human ingenuity in dealing with the unanticipated changes in tastes and demands.

Keywords: Artificial intelligence, Data, Manufacturing, Machine learning, Robotics.

INTRODUCTION

Artificial intelligence (AI) is a technology that accentuates the formation of smart and intelligent machines that work and respond like individuals. The machines or smart computer with the capability to accomplish intellectual roles such as observing, learning, decision making, reasoning, and solving problems are called artificial intelligence computers or machines. Whereas, the robotics is a field in artificial intelligence which deals in creating smart and intelligent robots using deep learning technology. Robotics technology is a field of artificial intelligence, which is comprised of all the Engineering fields such as mechanical, electrical and computer engineering for planning, designing, assembly, development and application of AI-based robots as shown in Fig. (**1**)

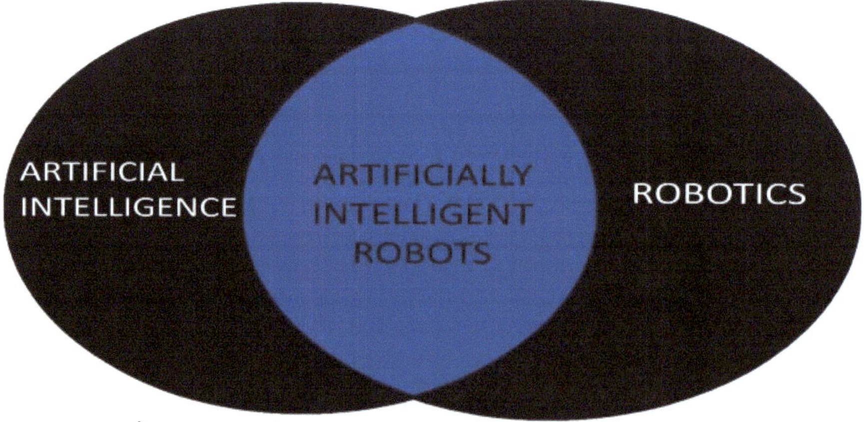

Fig. (1). Representation of AI robotics with Venn Diagram.

There are the smart machines working in real-time [1, 4, 6]. They are positioned for monitoring the objects by detecting, accumulating, shifting, modifying the basic properties of the object, damaging it, or to have an intelligent impact thus releasing manpower from doing a monotonous job without getting tired, or disturbed. The AI robots have advanced mechanical structure created to achieve a task with high sensitivity and accuracy as shown in Fig. (**1**). It also comprises electrical modules that power as well as control the mechanical structure of robots. Furthermore, it contains some advance level of software programming that governs, in what manner a robot does multiple tasks simultaneously.AI and robotics are the intelligence revealed by smart machine and robots to accomplish tasks that usually involve intelligence, human knowledge, and learning, skills, such as perceiving, learning, reasoning, solve problems, decision making and visualizing, *etc.* The emerging technologies comprise a variety of subset technologies like machine learning and deep learning, which are be learned with time, as subject to more and more data as shown in Fig. (**2**).

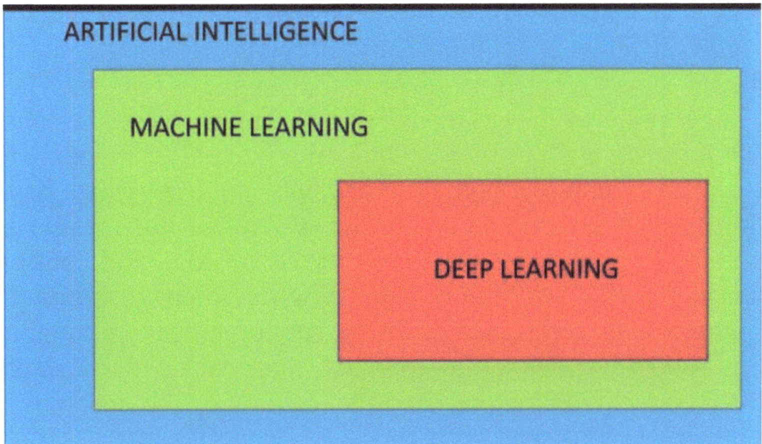

Fig. (2). Representation of AI and its fields.

Machine learning is the ability of machines or computers to learn without being programmed. It uses numerous algorithms to analyze and evaluate data, after analyzing the data, learn from it, and then make judgments or deliver an output. Therefore, machine learning algorithms permit systems to execute a task by training the machines by applying accessible data to inputs. On the other hand, deep learning is a subset of AI and machine learning, which are based on biological neural network learning representations for solving complex problems as humans do. Learning in AI can be supervised type, semi-supervised, reinforced or unsupervised type [2]. Supervised learning as the name indicates the presence of a supervisor as a teacher. supervised learning is a learning in which the teaching to the machine is done by labeled data, which indicates certain data is previously marked with the correct solution. Later Than, the machine is presented with a fresh set of instances(data). Consequently, the algorithm based on a supervised approach analyses the training data set and generates an appropriate conclusion from the learning of previously labeled data. For example, assume a basket filled with various types of fruits. The initial stage is to train the system with all dissimilar fruits in sequential order. If the shape of an item is rounded and imprint at topmost is having Red color, then the fruit will be labeled as Apple. On the other hand, an object is a long curving cylinder with Green-Yellow color, so it will be labeled as a fruit named Banana. Let us assume after training from the labeled data if a machine has been input with a new similar sample of fruit from the basket and asked to detect it. Because the system has previously learned from earlier data. The machine gives an accurate output. Supervised learning classified into two categories of algorithms such as Classification and Regression. On the other hand, unsupervised learning is done with data that is neither labeled nor classified. Thus, it is allowing the algorithm to act on that information without any guidance. Now, the machine must group unsorted data according to patterns, similarities, and differences without any previous training of data. Unlike supervised learning, no teacher is present that suggests, no such teaching or training has been given to the system or machine. Therefore, the unsupervised based machine is constrained to find the hidden arrangement in unlabeled data by itself. Let us take an example of both dogs and cats' pictures, which have not been trained previously with any data set. Now, the system has no clue about the shape of dogs and cats. Hence, it can't be categorized in dogs and cats or animals. Although, unsupervised algorithms can classify them according to their similarities, patterns, and differences. Finally, reinforcement learning, in artificial intelligence, is following a dynamic approach that teaches the machine using a reward and punishment technique. In this technique, the machine learns by interacting with its environment [3, 4].

ARTIFICIAL INTELLIGENCE AND ROBOTICS IN MANUFACTURING

The manufacturing industry has been considered with the highest degree of self-learning automation, complete technically AI robotic factories still seemed far away. Still, AI-characterized robotics is placed to shift that trend. The AI-enabled robots with superior ingenuity and self-sufficient learning abilities will transform the production process in Industries [5 - 8]. Whereas, research in this sector is facing major challenges, in terms of budget constraints. To overcome these designs, parameters for raw material, manufacturing techniques, and other details can be integrated into the software which performs on all feasible arrangements of input variables and recommends best designs. The output is further enhanced and improved with each repetition with the utilizing of artificial intelligence techniques. Hence, these promising technologies can offer us with superior designs and wipe out the trash. Currently, fault detection methods and tools are not accurate as compared to AI-based machines, resulting in a worthless production of trillion rupees in the manufacturing sector. To beat this a simulated model of a procedure, product, and supply chain can be formed called a digital twin. It is a blend of data with intelligence that signifies the behavior and framework of a system. This technique can improve the efficiency of data usage by handling only crucial information and removing junk information. The machine learning algorithms have an advantage in visual pattern identification, assisting in physical scrutiny of assets in the complete supply chain management [9 - 11]. Using supervised or unsupervised algorithms, applications can distinguish with damage or wear and tear. It can be able to organize the list of actions and recommend a corrective solution for restoring the problem in assets. Supply Chain Management can be done by advanced AI algorithms. As the supply chain is the backbone of the production industry. AI-enabled machines or robots will help in reducing cost in transportation, management of warehouse and efficient supply chain management. These machine learning algorithms are capable of processing huge, diverse data set on a real-time basis for demand-supply prediction. In an AI-based supply chain model, reduction in freight costs and improving supplier delivery will be accomplished [12]. Moreover, the process of inventory management can be accomplished with AI-enabled machines in the future. Because, most of the supply chains have to face the recent face of globalization, unstable demand, and progressively distinguished, adapted, and composite products. There is a massive task of matching customer facilities while taking care of inventories. The discomfort zones in inventory management have always been overstocking of inventory and failure to track important things or items. Here, AI machines will help in real-time supervising and optimization of assets. AI-enabled robots can assist with improved supply and demand planning. The foremost purpose of an inventory optimization result is to deliver customers with the precise product, at the accurate location, the right time along with the lowest inventory management. There will be a smart optimization solution that routinely modifies inventory levels to precisely meet up customer demands, enhancing service levels, reducing costs, and boosting inventory turnover of companies [13, 14]. Machine learning algorithms will help in inventory optimization across the whole supply chain and increase protection by automating hazardous activities. Processing real-time data and dangerous jobs can be handled with minimum impact with the deployment of AI robotics. These AI robots can work in risky situations like furnaces, hazardous material, boilers, welding, and die-castings, *etc.* while the above at a precision level cannot be achieved by existing procedures. Furthermore, advancements in the field of emerging technology like AI robotics is opening streets for a hybrid workforce where robots and machine manage collectively. They both work together to improve productivity along with creating additional posts that are facilitated by AI technology. Nowadays, robots are doing work in a fashion as illustrated in science fiction stories before. Although there are immense scopes and advantages of using robotics enabled with AI in manufacturing, the benefits of AI robotics far surpass the challenges in its implementation. The scope and potential of this technology in manufacturing and production will make the next industrial revolution in the future [15 - 18].

OPPORTUNITIES AND ADVANTAGES

The employ of AI and robotics will boost productivity and can create more manufacturing production work in countries like India. Labor is expected to get a substantial portion of the profits with AI and robotics technology. The implementation of AI and robots is exceptionally helpful in the manufacturing sector as it will improve the quantity and quality of productions, efficiency and speed in production. The induction of AI robots is to turn out to be a boom in the manufacturing industry [19]. AI robots can take training with labeled data set like humans learning, to operate more efficiently, including the new instructions that were not initially projected in the initial programming.AI based Robots will work as human do, but in spite of break, they will work continuously, As they don't get bored, hungry or tired. The abilities of robotic with AI technology are far and wide beyond human capability, once it extends to miniaturization and precision measurements [20]. Moreover, these emerging technology in the manufacturing sector will bring exceptional superior quality guarantees in the future. AI and robotics can transform the complicated task into consistent along with precise solutions. With the increasing operation of the industrial robot has drastically marked an increased deployment of AI in manufacturing processes. Artificial Intelligence (AI) is widely applied and is gradually approaching in the manufacturing segment, easing the automation with learning and error-free decision-making using learning algorithms. AI-driven systems are placing a simpler route to the future manufacturing industry by producing a lot of benefits by presenting innovative prospects, boosting production of goods with improving efficiencies, by implementing general AI to make machine interaction closer to human interaction. The AI-enabled robots with superior ingenuity and self-sufficient learning abilities will transform the production process in Industries. Whereas, research in this sector is facing major challenges in terms of budget constraints. To overcome these designs, parameters for raw material, manufacturing techniques, and other details can be integrated into the software which performs on all feasible arrangements of input variables and recommends best designs [21]. The output is further enhanced and improved with each repetition with the utilization of artificial intelligence techniques. There are several advantages with AI and robotics in the manufacturing sector as shown in

Fig. (**3**). The detailed benefits of AI based systems are given in detail.

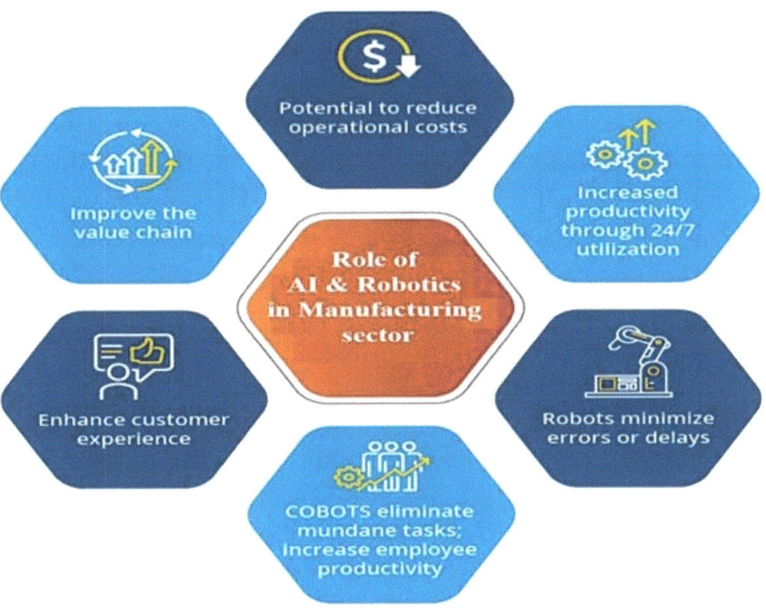

Fig. (3). Role of AI and Robotics in Manufacturing Industry.

a). Direct Automation

The introduction of robotics and artificial intelligence with big data analytics in heavy industry and the installation of new machines by making them programmable with a logic controller. The emerging technologies will Increase the use of precision sensory equipment in the future industries. This process incorporates information generation, recording, and analysis for all phases of the manufacturing process, encompassing everything from temperature sensing to product selection, picking, and packaging. SCADA (supervisory control and data acquisition) with deep learning can react automatically to the fluently produced data and adjust the tiniest events without no option to individual involvement. The robotics and big data analytics processed by artificial intelligence can significantly enhance implementation and performance throughout the entire manufacturing process and the whole process can be controlled remotely [22].

b). 24/7 Production

Human beings are creatures that require routine attention such as care, rest, water and food to work limited time in a day. However, for every manufacturing unit to work continuously, it is essential to initiate shifts, employing four workers round the clock to complete one cycle of 24 hours. The advantages of AI robots include no breaks, no fatigue, and regular working hours without interruption. Moreover, they are efficient in doing work in the manufacturing unit round the clock. This results in the growth of manufacturing facilities, which is gradually essential to meet up the requirements of global and local customers. Additionally, AI robots are more effective in many fields like an assembly line, selecting, picking of items and packaging. As a result, they can significantly slash turn-round times in various departments of the manufacturing sector [18 - 21].

c). Safety

The humans creatures are questionable and inclined to making botches, particularly if they are tired or diverted. Mistakes and mischances do happen on the plant floor and in any development or handling environment; a propensity which can be all but eliminated by AI and robotics. Inaccessibility get to control implies a decrease in human assets, particularly when they work is unsafe or requires superhuman exertion. Indeed, customary working situations will cut down on the frequency of mechanical mischances and lead to a by and large enhancement in security. In expansion, more progressed tangible hardware coordinates with IIoT gadgets make the establishment of security watches and obstructions a less difficult and more viable degree to ensure human lives [5].

d). Lower Operational Costs

Numerous companies are seeing the presentation of AI into the fabricating industry with fear because it requires a gigantic capital venture. On the other hand, the ROI is critical and increments as time go on. Once intelligent machines start to require over the day by day exercises of a production line floor, businesses will advantage through impressively decreased working costs, with prescient support making a difference moreover to decrease machine downtime. It is getting to be easier and cheaper to meet these needs and utilizing AI and ML methods implies that the entire generation prepare will be more cost-effective. Joining machine learning and CAD implies that frameworks can be planned and tried in a virtual demonstrate sometime recently they are put into generation, in this way diminishing the fetched of trial-and-error machine testing [23].

e). Greater Efficiency

With the help of emerging technologies such as big data analytics processed with

Ai algorithms the overall trends of an industrial sector can be forecast, patterns recognition with training and market developments prediction with time, macroeconomic cycles and even weather patterns with the help of machine learning and data analytics. AI has the capacity with machine learning to expect data, to refine forms, and to track incoherencies, all the way down the supply chain from source to wrapped up item. This can be especially supported by such advances as RFID following, which empowers materials to be followed without having to go through a physical prepare such as a bar code per user [16, 19].

f). Quality Control

AI robotics is additionally greatly valuable for carrying out prescient maintenance on machines, apparatus and other hardware in industries. Furthermore, utilizing various sensors with inbuilt AI techniques to track execution and working conditions of machines to check, predict all types of failures, and take activity to cure them before they happen. This could result in quicker feedback, with reduced downtime of the machine. Hence, it results in overall efficiency and productivity. They can also identify the nano or microscopic faults, examining them at resolutions far beyond the capability of normal individual imagination, consequently enhancing productivity and boosting the number of products that will by-pass quality checking and control department. AI-enabled robotics makes a big difference to pace up several plan models and makes advances precision to a massive scale. This prevents the requirement for quality control and in-process assessment by human creatures, which is time-consuming and frequently fallible [22, 23].

g). Quick Decision Making

AI robotics can offer quick decision making in the manufacturing sector by computerizing complex tasks that would take the only nanosecond. Whereas, it takes too much time with human help. It can moreover see information more deeply and at information to discover patterns that people may never capture. The utilize of manufactured insights in decision-making must fit workflows and designs that make sense for clients. AI incorporates the computerization of cognitive and physical assignments. It makes a difference individual perform tasks quicker and way better and make superior choices. It empowers the robotization of choice-making without human intercession. AI can improve robotization in this way decreasing seriously human labor and repetitive tasks [16 - 18]. Therefore, there are numerous more ways in which AI is making a distinction. The detail of decision making or another task by AI has been illustrated in Table **1**.

Table 1. AI decision making with description and examples.

Type of Decision	Description	Examples
Classification	Categorize as good or bad	Potential borrower likely to charge-off, the customer likely to be lost, object in the image likely to be a cat, *etc.*
Grouping	Detect natural grouping of customers	The Segmentation of customers into the clusters
Association	Find meaningful correlations between various entities	Suggest the next best product or offer based on other related customers
Prediction	Predict the future price of a continuous quantity	Forecast deals, forecast price of customers over the lifetime
Natural Language Processing	Identify structure, recognize entities, extract meaning	Translation, generate summary from details, identify entities and their relationships
Visual Perception	Identify objects in an image or video, generate the 3D world	Understand edges, texture, objects. Reconstruct 3D world, detect anomalies
Robotics	Manipulate objects in the real world	Mobile manipulators such as self-driving cars, assembly line part picking robots

CHALLENGES OF AI AND ROBOTICS IN MANUFACTURING

Implementing AI and robotics in industries is not as easy as procuring the robot and plugging for the application There are numerous different financial and government policies considerations to consider before an AI can constructively automate anything. Each producer will have one of a kind of challenges based on their industry and their planning along with the business model. However, there are typically numerous challenges in implementing AI for manufacturing. The challenges of implementing AI and robotics to open the esteem lies within the change of raw information to intelligent predictions for quick decision-making. In general, there are various challenges in the implementation of artificial intelligence [25].

a). Data

Designing frameworks presently create a part of the information and cutting-edge industry is a huge information environment. In any case, mechanical information, as a rule, is organized but it may be of poor quality. The quality of the information may be destitute, and not at all like other consumer-faced applications, information from mechanical frameworks often have clear physical implications, which makes it harder to compensate for the quality with volume. Information collected for preparing machine learning models, as a rule, is missing a comprehensive set of working conditions and wellbeing states/fault modes, which may cause untrue positives and untrue negatives in online usage of AI frameworks. Mechanical information designs can be exceedingly temporal and deciphering them requires space mastery, which can barely be saddled by simply mining numeric information [26].

b). Speed

Manufacturing procedure occurs quickly, and the machines/apparatus and other types of equipment can be high-priced, the AI along with data analytics approach for various applications must be employed in real-time to detect irregularities instantly to prevent trash and other effects. Cloud-based solutions may be effective and rapid techniques, but not suited for specific computation efficiency conditions [24 - 26].

c). High Fidelity Requirement

Unlike customer-encountered AI-based recommendation systems that incorporate a superior tolerance for errors, still a minimal rate of processing the errors may price the overall reliability of robotics and AI systems. Engineering AI-enabled applications are typically dealing with crucial problems related to protection, consistency, and operations. Any breakdown in projections might incur adverse economic impact on the customers and discourage them from trusting on AI-enabled systems.

d). Interpretability

The general AI-enabled systems should also go ahead of prediction results and provide root cause evaluation for glitches. This involves data scientists who must work with field specialists during the development of generalized application and it must incorporate field process detail into the modeling process.

e). Employee Skillset and Training

When new AI robotic techniques are executed, a fresh degree of knowledge is expected from the employees. Numerous employees will have to be taught on how to manage in this different atmosphere, while other workers will have to be appointed that have the appropriate qualifications, learning, and experience in AI and robotics.

f). Safety Measures

AI and robotics present numerous modern security risks into the working environment and there are strict controls, as well as firm punishments, encompassing automated security. Producers must get ready for this sometime recently the robots are introduced to guarantee compliance and make a secure environment for experts.

g). Budgeting for the Cost of the System

Manufacturers may face functioning and monetary challenges while implementing AI-enabled applications/software into the workplace. Such applications/software be able to contain something from trustworthy and super-pace Internet too expensive machinery. There is usually a large investment related to AI and robotics, though prices are firmly dropping. But still the parameter a huge challenge for existing players to update their existing system. The AI and robotics technology into the manufacturing industry would necessitate a huge capital investment for startups and existing manufacturers. Hence, sales and production of goods numbers should remain stable during the projected ROI time to compensate for the initial time of investment [27].

h). Managing Product Workflow

There are numerous contemplations when deciding what item workflows will see like when an AI is introduced. The introduction and speed of portion introduction to the AI robotic must be calculated to guarantee the greatest efficiency without creating much as well for existing frameworks to handle. Representative skillsets, unused security conventions, budgeting, and overseeing item workflows are a few of the greatest challenge's producers confront when executing mechanical automated mechanization frameworks. There's distant more to realizing emerging technologies than acquiring and buying them. Manufacturers ought to be ready for the challenges to get the maximum out of their AI-enabled robotization frameworks [24 - 27].

Now, it's time to convert the conventional fabricating plant into a keen AI-enabled plant, making utilization of the AI, Web of Things (IoT), data analytics and Cloud computing because it does. As ever, there's a challenge – and in this case, it's one of fetch. New era will experience lot of changes and new research experience for researcher, industries and academicians [28 - 32].

CONCLUSION

The manufacturing industry is ideally suited for the application of artificial intelligence and robotics. Even though the AI revolution is yet in its initial stages, we are now observing substantial benefits from this technology. The use of AI from the primary design procedure, production floor, full supply chain and complete administration, artificial intelligence is intended to modify the manner in manufacturing of products and incorporation of process materials ever. Both AI & robotics can be embedded to present manufactured goods and applications to create more efficient, consistent, safer, and robust in terms of life and operation. Together With the support of AI and robotics, the level and speed of industrial automation have been profoundly changed.AI enabled technologies to enhance the execution and develop the capacity of conventional applications. An instance is AI-enabled collaborative robots. The robotic arms can learn the movement and route exhibited by human operators and execute a similar job after supervised training which is a part of artificial intelligence. In addition to this, AI also automates the method applied to involve human involvement.

CONSENT FOR PUBLICATION

Not applicable.

CONFLICT OF INTEREST

The authors confirm that this chapter content has no conflict of interest.

ACKNOWLEDGEMENTS

The authors would like to express their sincere thanks to the editor and anonymous reviewers for their time and valuable suggestions.

REFERENCES

[1]　X. Shang, X. Liu, G. Xiong, C. Cheng, Y. Ma, and T.R. Nyberg, "Social manufacturing cloud service platform for the mass customization in apparel industry", *Proc. IEEE Int. Conf. Service Oper. Logistics Informat.,* pp. 220-224, 2013.
[http://dx.doi.org/10.1109/SOLI.2013.6611413]

[2]　J.H. Thrall, X. Li, Q. Li, C. Cruz, S. Do, K. Dreyer, and J. Brink, "Artificial intelligence and machine learning in radiology: Opportunities challenges pitfalls and criteria for success", *J. Am. Coll. Radiol.,* vol. 15, no. 3 Pt B, pp. 504-508, 2018.
[http://dx.doi.org/10.1016/j.jacr.2017.12.026] [PMID: 29402533]

[3]　X. Xu, and Q. Hua, "Industrial big data analysis in smart factory: Current status and research strategies", *IEEE Access,* vol. 5, pp. 17543-17551, 2017.
[http://dx.doi.org/10.1109/ACCESS.2017.2741105]

[4]　Z.X. Guo, W.K. Wong, S.Y.S. Leung, and M. Li, "Applications of artificial intelligence in the apparel industry: A review", *Text. Res. J.,* vol. 81, no. 18, pp. 1871-1892, 2011.

[http://dx.doi.org/10.1177/0040517511411968]

[5]   R. Nayak, and R. Padhye, *"Artificial intelligence and its application in the apparel industry" in Automation in Garment Manufacturing.* Elsevier: Amsterdam, The Netherlands, 2018, pp. 109-138.

[6]   C.K.H. Lee, K.L. Choy, K.M.Y. Law, and G.T.S. Ho, "An intelligent system for production resources planning in Hong Kong garment industry", *Proc. IEEE Int. Conf. Ind. Eng. Eng. Manage.,* pp. 889-893, 2012.

[7]   G.M. Nasira, and P. Banumathi, "An intelligent system for automatic fabric inspection", *Asian J. Inf. Technol.,* vol. 13, no. 6, pp. 308-312, 2014.

[8]   M-K. Chen, Y-H. Wang, and T-Y. Hung, "Establishing an order allocation decision support system *via* learning curve model for apparel logistics", *J. Ind. Prod. Eng.,* vol. 31, no. 5, pp. 274-285, 2014.

[9]   G. Baryannis, S. Dani, and G. Antoniou, "Predicting supply chain risks using machine learning: The trade-off between performance and interpretability", *Future Gener. Comput. Syst.,* vol. 101, pp. 993-1004, 2019.
        [http://dx.doi.org/10.1016/j.future.2019.07.059]

[10]   R. Sharma, S.S. Kamble, A. Gunasekaran, V. Kumar, and A. Kumar, "A systematic literature review on machine learning applications for sustainable agriculture supply chain performance", *Comput. Oper. Res.,* vol. 119, 2020.104926
        [http://dx.doi.org/10.1016/j.cor.2020.104926]

[11]   R. Stuart, and P. Norvig, *Artificial Intelligence: A Modern Approach.* Prentice-Hall: Upper Saddle River, NJ, USA, 2009.

[12]   M. Mohri, A. Rostamizadeh, and A. Talwalkar, *Foundations of Machine Learning.* MIT Press: Cambridge, MA, USA, 2012.

[13]   S.V. Kumar, and S. Poonkuzhali, "Improvising the sales of garments by forecasting market trends using data mining techniques", *Int. J. Pure Appl. Math.,* vol. 119, no. 7, pp. 797-805, 2018.

[14]   H. Pahuja, M. Tyagi, S. Panday, and B. Singh, "A novel single-ended 9T FinFET sub-threshold SRAM cell with high operating margins and low write power for low voltage operations", *Integration (Amst.),* vol. 60, pp. 99-116, 2018.
        [http://dx.doi.org/10.1016/j.vlsi.2017.08.004]

[15]   H. Pahuja, M. Tyagi, B. Singh, and S. Panday, "leakage immune single ended 8t sram cell for ultra-low power memory design", *Journal of Engineering Science and Technology,* vol. 14, no. 2, pp. 629-645, 2019.

[16]   Y.H. Pan, "Heading toward artificial intelligence 2.0", *Engineering,* vol. 2, no. 4, pp. 409-413, 2016.
        [http://dx.doi.org/10.1016/J.ENG.2016.04.018]

[17]   B-R. Li, Y. Wang, and K-S. Wang, "A novel method for the evaluation of fashion product design based on data mining", *Adv. Manuf.,* vol. 5, no. 4, pp. 370-376, 2017.
        [http://dx.doi.org/10.1007/s40436-017-0201-x]

[18]   S-W. Hsiao, C-H. Lee, R-Q. Chen, and C-H. Yen, "An intelligent system for fashion colour prediction based on fuzzy C-means and gray theory", *Color Res. Appl.,* vol. 42, no. 2, pp. 273-285, 2017.
        [http://dx.doi.org/10.1002/col.22057]

[19]   M. Mohri, A. Rostamizadeh, and A. Talwalkar, *Foundations of Machine Learning.* MIT Press: Cambridge, MA, USA, 2012.

[20]   R. Stuart, and P. Norvig, *Artificial Intelligence: A Modern Approach.* Prentice-Hall: Upper Saddle River, NJ, USA, 2009.

[21]   P. Yildirim, D. Birant, and T. Alpyildiz, "Data mining and machine learning in textile industry", *Wiley Interdiscip. Rev. Data Min. Knowl. Discov.,* vol. 8, no. 1, 2018.e1228
        [http://dx.doi.org/10.1002/widm.1228]

[22] S.K. Acharya, "Singh, V. Pereira, P. Singh, "Big data knowledge co-creation and decision making in fashion industry", *Int. J. Inf. Manage.*, vol. 42, pp. 90-101, 2018.
[http://dx.doi.org/10.1016/j.ijinfomgt.2018.06.008]

[23] J. Lee, E. Lapira, B. Bagheri, and H.A. Kao, "Recent advances and trends in predictive manufacturing systems in big data environment", *Manuf. Lett.*, vol. 1, no. 1, pp. 38-41, 2013.
[http://dx.doi.org/10.1016/j.mfglet.2013.09.005]

[24] K. Lee, *Artificial intelligence, automation, and the economy.* The White. House Blog, 2016.

[25] H. Leurent, and E.D. Boer, "Fourth Industrial Revolution Beacons of Technology and Innovation in Manufacturing", *White Paper,* January 2019.10https://www.weforum.org/whitepapers/fourth-industrial-revolution-beacons-of-technology-andinnovation-in-manufacturing

[26] J. Lee, M. Azamfar, and J. Singh, "A Blockchain Enabled Cyber-Physical System Architecture for Industry 4.0 Manufacturing Systems", *Manuf. Lett.*, 2019.
[http://dx.doi.org/10.1016/j.mfglet.2019.05.003]

[27] H. Xiao, B. Muthu, and S. Kadry, "Artificial intelligence with robotics for advanced manufacturing industry using robot-assisted mixed-integer programming model", *Intell. Serv. Robot.*, 2020.
[http://dx.doi.org/10.1007/s11370-020-00326-7]

[28] C. Chow, and C. Leung, Technology Challenges
[http://dx.doi.org/10.2174/9781681087498118010025]

[29] J. Jiang, Z. Min, X. Ma, Y. Zhang, and S. Song, "Application of Robot to Hand Function Rehabilitation", *Recent Pat. Mech. Eng.*, vol. 11, no. 1, pp. 2-14, 2018.
[http://dx.doi.org/10.2174/2212797611666180118094318]

[30] C. Chow, and C. Leung, The Fourth Industrial Revolution
[http://dx.doi.org/10.2174/9781681087498118010017]

[31] J. Pérez-Sianes, H. Pérez-Sánchez, and F. Díaz, "Virtual Screening Meets Deep Learning", *Curr Comput Aided Drug Des,* vol. 15, no. 1, pp. 6-28, 2019.
[http://dx.doi.org/10.2174/1573409914666181018141602] [PMID: 30338743]

[32] M.T. Almansoori, X. Li, and L. Zheng, "A Brief Review on E-skin and its Multifunctional Sensing Applications", *Curr. Smart Mater.*, vol. 4, no. 1, pp. 3-14, 2019.
[http://dx.doi.org/10.2174/2405465804666190313154903]

SUBJECT INDEX

A

Abrasive erosion process 63
Abrasive Air Jet Machining (AAJM) 72, 74
Abrasive Fluid Jet Polishing (AFJP) 83
Abrasive Grains 1
Abrasive Jet Machining 18, 45, 71, 73, 83
Abrasive Material 81, 82, 83, 59
Abrasive Size 79, 80, 81
Abrasive Water jet 18, 33, 36, 51, 58, 73, 74, 75, 97
Abrasive Water Jet Machining 18, 33, 36, 51, 58, 73, 74, 75, 97,
Abrasive Wire EDM 109
Alloy steel 29, 58, 113
Artificial intelligence 133, 134, 135, 138, 139, 145
Artificial Neural Network (ANN) 4, 5, 48, 53, 103, 112, 113

C

cemented carbides 95
Ceramic matrix 7
Chipping 6, 8
Chips 2, 49, 54, 72, 79
cobalt-bonded tungsten carbide 97
Complex Shape 29, 33, 35, 36, 42, 88
Composite Material 67, 76, 88, 92, 94, 95, 96
conductive materials 19, 90, 94
Conventional Machining 1, 2, 5, 9, 12, 21, 29, 44, 63, 77, 92, 103, 106
Coolant 12, 53
Crack Propagation 78, 110
Crater 20, 21, 33, 95, 96, 110
Cryogenic 72, 74, 83, 112
Cutting Forces 2, 3, 5, 6, 7, 8, 9, 12, 47, 50, 60, 61, 106
Cutting Tool 1, 10, 21, 23, 54, 92
Cutting Velocity(CV) 107, 123, 124

D

De-ionized water 106
Design of Experiment (DOE) 47, 53, 103
Dielectric Fluid 19, 20, 32, 104, 106, 127, 128, 129
Dielectric Pressure 27, 124
dimensional accuracy 33, 34, 35, 36, 104, 110, 115,
Ductile materials 62, 54

E

EDM Electrode 26, 28, 33
Elastic Erosion 71, 78
Electric pulses 122
Electrical Conductivity 17, 21, 22, 24, 25, 29, 90, 91, 92, 111
Electrical Discharge Machining (EDM) 17, 18, 45, 47, 88, 90, 93, 104
Electron Back Scatter Diffraction (EBSD) 103, 113
Elongation 108
Energy dispersive spectroscopy (EDS) 103, 110

F

Feed Rate 3, 6, 8, 62, 81, 111, 114, 115
Flow Rate 61, 66, 71, 72, 77, 80, 81, 82, 83, 96
Flow Velocity Distribution 74
Functionally Graded Materials(FGM) 80

H

Hard and Brittle Material 1, 2, 5, 9, 12, 13
Heat Energy 20, 42, 43, 88
High pressure liquid pump(HPLP) 59
High speed water jet 59
Hybrid composite material 94
Hybrid Machining 1, 3, 5, 6, 11, 42, 47, 53